THEORIES AT WORK

THEORIES
AT WORK

On the structure and functioning
of theories in science,
in particular during the
Copernican Revolution

Marinus Dirk Stafleu

UNIVERSITY
PRESS OF
AMERICA

LANHAM/NEWYORK/LONDON

British Cataloging in Publication Information Available

Co-published by arrangement with the
Institute for Christian Studies, Toronto, Ontario, Canada

Cover/book design: Willem Hart
Typeset by Carroll Guen, Willem Hart Art & Design, Inc. Toronto

ISBN: 0-8191-6571-9 (pbk. alk. paper)
ISBN: 0-8191-6570-0 (alk. paper)

Acknowledgements

This book has grown out of a course in philosophy and history of science given at Utrecht, Netherlands, since 1982, and at Toronto, Canada, in 1985. I thank all participants in these courses for their critical comments.

I wish to express my deep gratitude to Dr. H. Hart and Dr. R. E. VanderVennen, Institute for Christian Studies, Toronto, Ontario; Dr. T. H. Leith, York University, Downsview, Ontario; and Dr. C. Jongsma, Dordt College, Sioux Center, Iowa, who read the manuscript and proposed many improvements in the text.

Critical comments on earlier versions of parts of the text were made by Dr. J.D. Dengerink and Dr. H.A.M. Snelders, State University, Utrecht; Dr. J. van der Hoeven, Free University, Amsterdam; and by my former colleague, Drs. J. ten Hoope.

Many thanks are due to Ms. Carroll Guen for typesetting the book.

I very much appreciate the cooperation of the Institute for Christian Studies, Toronto, under whose auspices this work is published.

M.D. Stafleu
Utrecht, The Netherlands, 1986

Contents

1. The logic of theories
    1.1. The Copernican revolution/...9
    1.2. The artificial character of a theory/...11
    1.3. The logical structure of a theory/...15
    1.4. Concepts/...19
    1.5. Statements and their context/...24

2. Explanation and prediction
    2.1. Physics and astronomy before Copernicus/...31
    2.2. Prediction/...38
    2.3. Explanation/...41
    2.4. Retrograde motion/...44
    2.5. The size of planetary orbits/...46

3. Principles of explanation
    3.1. The harmony of the spheres/...52
    3.2. Explanation of change in Aristotelian philosophy/...56
    3.3. Galileo on motion/...61
    3.4. Cartesian physics/...66
    3.5. Kepler's defection/...71
    3.6. Newton's dynamics/...75

4. The solution of problems
    4.1. The various functions of the statements in a theory/...84
    4.2. Normal science/...89
    4.3. The generation of problems/...95
    4.4. Axioms and the aim of science/...98
    4.5. Crisis and revolution/...103

5. The systematization of knowledge
    5.1. The reduction of theories/...108
    5.2. Putting theories to test/...112
    5.3. The sources of data/...117
    5.4. Objectivity/...125

6. Heuristics
    6.1. Induction/...129
    6.2. The method of successive approximation/...133
    6.3. The method of analogy/...139
    6.4. The method of mathematization/...143
    6.5. The function of technology/...147
    6.6. Theory of the opening process/...151

7. The principle of clarity
    7.1. Communication/...158
    7.2. Didactics/...162
    7.3. Polemic/...165

8. Science and society
    8.1. The organization of science/...171
    8.2. The emancipation of science/...174
    8.3. Science and the church/...177
    8.4. The social responsibility of scientists/...182

9. Parsimony and harmony
    9.1. The economics of science/...186
    9.2. The simplicity of the Copernican system/...189
    9.3. Copernicus' and Newton's systems of the world/...193
    9.4. The principle of harmony/...196

10. Criticism
    10.1. Popper on criticism/...200
    10.2. Immanent, transcendent, and transcendental
        critique/...205
    10.3. The critical function of the scientific community/...210

11. Commitment

    11.1. "On the hypotheses of this work"/...216

    11.2. The hypotheses of Copernicus and Tycho/...220

    11.3. Descartes on hypotheses/...223

    11.4. Hypotheses non fingo/...226

12. Belief

    12.1. The search for certainty/...232

    12.2. The status of natural laws/...238

    12.3. World views/...242

    12.4. The hypostatization of theoretical thought/...249

    12.5. The unity of the creation/...256

Notes/...261

Index/...288

1. The logic of theories

1.1. The Copernican revolution

In the introductory Chapter 1 a synopsis will be given of what we consider to be a *theory*, together with some examples taken from the Copernican era. For various reasons, the time-honoured expression "Copernican Revolution" has a strong appeal. It clearly points to the title of Copernicus' epoch making book, *De Revolutionibus Orbium Coelestium* (On the revolutions of the celestial spheres, 1543).[1] But its author, the Polish canon Nicolas Koppernigk, did not intend to start anything like what is now understood by a "revolution." The term "Copernican revolution" was probably first used in 1787 by the German philosopher Immanuel Kant, who coined it to emphasize a radically new point of view in his epistemology.[2] By implication, Kant recognized the revolutionary character of Copernicus' heliocentric theory, more than Copernicus himself did.

Like Thomas Kuhn in his book *The Copernican Revolution* (1957), we shall apply the expression "Copernican revolution" exclusively to the historical period from 1543 to 1687.[3] The latter year witnessed the publication of Sir Isaac Newton's *Philosophiae Naturalis Principia Mathematica* (Mathematical Principles of Natural Philosophy), which book was considered to contain the final and decisive proof of the Copernican theory.

The division of history into well-defined eras is unavoidably arbitrary. In science, the "Renaissance" is supposed to indicate the period between *c.* 1450 and *c.* 1620. It is characterized by its criticism of medieval science, and its call to return to ancient, especially Platonic, views. It was succeeded by "classical physics," dominated by mechanics, which period found its end about 1900. Copernicus and Tycho Brahe figure as typical Renaissance scholars, whereas René Descartes, Christiaan Huygens, and Newton are considered classical physicists, adhering to the new "mechanical philosophy." Johannes Kepler and Galileo Galilei started their careers as Renaissance scientists, but their most mature work is "classical" in spirit. They "crossed the watershed" between ancient and modern science, according to A. Koestler.[4] In several respects, Kepler's and Galileo's works form a turning point in the Copernican revolution, which comprises the final part of the Renaissance and the beginning of classical physics. Indeed, the Copernican ideology of the moving earth was the motor of the transition from Renaissance to classical physics. Copernicus started it, Kepler and Galileo were its chief advocates, and Newton brought it to completion. We call this period "Copernican" because almost all its heroes considered themselves "Copernicans", and their common creed was that the earth moves.

The Copernican revolution concerned astronomy and physics, mechanics, magnetism and optics. Simultaneously it saw a battle between at least four philosophies. Christianized during the middle ages, *Aristotelian* philosophy dominated the universities, and was mostly defended by conservative professors. It elicited the *Platonic-Pythagorean* reaction, the Renaissance philosophy, with its appeal to return to the Greek and biblical sources of civilization, unpolluted by medieval corruption. Up till Galileo, most Copernicans were under the spell of this philosophy. Thirdly, the *mechanical* philosophy imperceptibly replaced the Platonic views. It was a reaction to Aristotelian philosophy as well. Its main spokesman was René Descartes, but he was neither the first nor the last mechanist. Isaac Beeckman, Christiaan Huygens, Robert Boyle, and Gottfried Leibniz adhered to it in various degrees. Finally, we encounter the beginning of *empiricism*, a new philosophy opposing the rationalistic trends of the former three. Traces of it can be found in Kepler, Galileo, and Huygens. Francis Bacon was its prophet, and Blaise Pascal, Boyle, and Newton propagated it.

Isaac Newton's *Principia* (1687) marks the end of the Copernican revolution, and the beginning of Newtonianism, a new era. Newtonianism started later than classical physics, and ended earlier, about 1850, when it was superseded by a short revival of Cartesian mechanism.

Our ensuing quite extensive discussion of the Copernican revolution should not deceive the reader to assume that this book is a historical treatise. We aim to discuss neither the history of science, nor the history of philosophy for its own sake. Our aim is to study the nature of theoretical thought, in particular the nature of scientific theories. But it makes little sense to give an exposition of a "theory of theories" without testing it against the

history of ideas, both in science and in philosophy. Two particular reasons can be offered for taking the Copernican episode as a
test case.

First, the modern view of science as well as the modern view
of theories was born during the Copernican revolution. It is the
revolutionary idea that the basic axioms of a theory need not be
evident. The dynamic motif of Copernicanism, the creed that the
earth moves, is counter-intuitive, notwithstanding the fact that
by an overwhelming indoctrination nowadays everyone is made to
believe it. According to the Aristotelian philosophy, prevalent
before the Copernican revolution, any explanation should start
from known and well-understood premises. Gradually, it began to
dawn that the most powerful use of theories is to start from the
unknown, because it allows one to find the unobservable laws of
nature. This view on theories is related to the equally modern
conception of science, *i.e.*, the exploration of the lawful structure
of reality.

Second, we observe that Kant's *Kritik der reinen Vernunft*
(Critique of pure reason, 1781, 1787), in which second edition he
coined the term "Copernican revolution", was a philosophical
reflection on Newton's physics and on its implications for the
theory of knowledge. Nineteenth-century Kantianism was challenged by positivism, which in several varieties dominated the
philosophy of science between *c.* 1920 and *c.* 1960.[5] About 1960
another revolution started, initiated by Karl Popper's *The Logic
of Scientific Discovery* (1959), and Thomas Kuhn's *The Structure
of Scientific Revolutions* (1962).[6] In the ensuing philosophical debates, the Copernican revolution was a much discussed topic.
Hence, the Copernican period will not only provide us with materials to test our own theory, but also with an entry to review contemporary philosophies of science.

Theories are found in books. Therefore, our empirical study of
theories is based on an investigation of the books constituting the
Copernican revolution, in particular the books written by Copernicus, Kepler, Galileo, Descartes, and Newton. But theories are
also used by people. If we want to study theories, we have to find
out how theories function in real life, in particular in science. This
is why we call this book *"Theories At Work"* — we shall discuss
theories as they function in their context.

1.2. The artificial character of a theory

Is it possible to distinguish theoretical thought from non-theoretical thought? Let us try to answer this question without discussing the far more difficult problem about the nature of thought itself. Natural, non-theoretical thought is spontaneous, is characterized by an immediate relation between the thinking man or woman, the *subject*, and the *object* of his or her thought. In theoretical thought this direct relation is interrupted, because man puts a theory between himself and his object of thought. A theory is like a medium, mediating between subject and object, it is an *instrument*. We shall elaborate this, without committing ourselves to instrumentalism.

Natural and artificial seeing

In order to make the instrumental character of theories clear, let us first consider a different kind of instrument, instruments to improve vision. *Seeing* is a natural activity of men, and of all animals having eyes. We see objects in our environment — a tree, a tower, a car. Occasionally, we also look at a picture of a tree. In an artificial manner, we see a tree, whereas in a natural manner, we see a picture. In a natural way, we cannot see our own face, or the phases of Venus. Using a mirror, we see naturally a picture, but artificially our own face. Using a telescope, we see naturally an image, but artificially the phases of Venus. Artificial seeing is not contrary to natural seeing, but depends on it.

We invent optical instruments in order to see better than would be possible in a natural way. In Holland, one cannot see the Eiffel tower directly. But one can see a picture of it, a photo, a miniature, or a TV-picture. This kind of artificial seeing seems to be an exclusively human activity. Animals do not see the Eiffel tower, if they see its picture — they only see a piece of paper. Nor do animals invent instruments to improve seeing.

Contrary to natural seeing, artificial seeing has a history. Medieval painting, Renaissance and modern art differ widely from each other. Photography was invented in the 19th, television in the 20th century. It is more than a coincidence that the telescope and the microscope were invented during the Copernican revolution. The new movement made these discoveries possible, and needed them all the same. The idea that instruments may be able to improve our experience was opposed by conservative Aristotelians, and furthered by progressive Copernicans. It is associated with the new insight that theories can be instrumental in the exploration of reality.

We know that Galileo was the first to use the telescope for astronomical observations — it is a historical event (1609). He discovered new stars, mountains on the moon, and Jupiter's satellites, besides Venus' phases and the sunspots. He used these discoveries in his propaganda for the Copernican theory.

Artificial seeing is natural seeing, opened up by historical, human activity. It is characterized by the use of artifacts: pictures like photographs and paintings, and instruments like mirrors and telescopes.

Natural and artificial thought

There is no need to *define* "seeing" — we know what it is. Similarly, natural thinking is familiar to us; it is a natural activity of men and women, and perhaps of all animals having brains. Natural thought concerns trees, towers, or stars, good or evil deeds, families and churches, colours and paintings.

All thought is characterized by dissociation and association, by making distinctions and connections, by *logic*. Thinking beings are logical sub-jects, and they think about logical ob-jects. The logical objects of natural thought are concrete, everyday things, events, and their relations. The thinker distinguishes and relates them.

A theory, too, is an object, though it is certainly not a thinking subject, as it does not think. But it is not a *logical* object. Except for philosophers, nobody thinks about theories. A theory is an *artifact*.[7] We make theories, invent them, improve them, and use them. Theories are objects, because they are *made* by people, not because people *think* about them. We use theories as instruments in our thought. Theoretical thinking is natural thinking, opened up by the use of instruments. We form concepts of concrete things, of events, and of relations, and we think theoretically about them.

Contrary to natural thought, theoretical thought has a history, the history of ideas. It is generally assumed that theoretical thought originated with the Greeks, about 600 B.C. The theories of the Greeks differ from the medieval theories, from those of the 17th century, and from ours.

Theoretical thought is concerned with statements, and statements concern concrete things, events, and relations. This constitutes a problem, the problem of the relation between artificially conceived theories, statements, and concepts on the one hand, and concrete things, events, and relations on the other hand. This problem is related to the fact that by using an instrument we enlarge our power of seeing or thinking, but simultaneously diminish our field of vision or attention. Using a microscope we can see much better than without, but at the same time we see much less. Moreover, we eliminate our other senses. If we look at a real dog, we not only see him, but we also smell him, and hear him. This continuity of sensory experience is interrupted if we look at a picture of a dog. Therefore, a dog can see another dog, but if we show him a picture of a dog, he does not recognize it. In theoretical thought we make abstractions, we restrict our conceptual activity, and we switch off our other modes of experience — for instance, our feelings.

Often, theoretical thought is at variance with natural thought. An example is the Copernican leading idea of the moving earth, which is counter-intuitive. It is "...almost against common sense to imagine some motion of the Earth".[8] Therefore, it is required that a theory be proved. Whether this is possible and to what extent will be discussed later on.

The three worlds of Karl Popper

The distinction between thinking subjects, objects to be thought about, and theoretical artifacts, more or less corresponds with Popper's distinction of "three worlds": "...the first is the physical world or the world of physical states; the second is the mental world or the world of mental states; and the third is the world of intelligibles, or of *ideas in the objective sense*; it is the world of theories in themselves, and their logical relations; of arguments in themselves; and of problem situations in themselves".[9]

This correspondence is not perfect, however. Popper seems to think that his classification is complete and exhaustive, that everything belongs to one of his worlds. Our classification is *merely logical*, it has a merely logical function. There are more subjects than logical subjects: mathematical subjects like numbers and spatial figures, physical subjects like atoms and stars, biological subjects like plants and animals. Even man is more than a thinking subject. He has feelings, he acts, he loves, he believes. Next, there are more objects than logical objects. The magnitude of a spatial figure, its length or volume, is a spatial object. A road is a kinematic object, it does not move, but is indispensable for traffic. Food is a biological object, it does not live, but is a condition for life. A painting is an object of art. There are also far more artifacts than theories and ideas — telescopes, houses, cars, clothing.

Hence, our distinguishing logical subjects, logical objects, and theories as logical instruments, has a functional character. It concerns one of our fundamental modes of experience. The logical aspect of experience is only one segment of human experience, and of being in general.[10] Other aspects, like the numerical, the spatial, the kinematic, and the physical, have played equally important parts in the history of the Copernican revolution, as we shall see in due course (see Chapter 3).

1.3. The logical structure of a theory

A human mode of experience and activity, logic implies making distinctions and connections. One of the most important logical distinctions to be made is that between the truth and falsity of statements. Hence, the most general logical function of a theory is to establish the truth content of a statement. Omitting statements which are more or less probable, we shall restrict ourselves to propositions which are either false or true.

Distinctions are made by people. As a logical subject, man is actively subject to logical laws, which he applies in his arguments, and which he has to obey if he wishes to argue rightly.

The most important logical law seems to be the *law of non-contradiction*. Within a certain context a statement and its negation cannot both be true. The qualification "within a certain context" is of crucial importance. It points out that theoretical reasoning has a *relative* value. A statement can be true in one context, false in another one. But a theory is inconsistent if it simultaneously contains a statement and its negation.

The logical definition of a theory
What is a theory? The Greek word *theoria* means something like "contemplation", but already the earliest Greek philosophers connected *theoria* with proof, or deductive reasoning. We shall take for granted that a theory invariably implies deduction. It is often assumed that a theory should start from well-known and accepted truths, in order to arrive at new statements or theorems. Other people maintain that scientific theories start from the unknown, from hypotheses which should explain the observable. In this case, theories are even identified with hypotheses,[11]which is deplorable, for hypotheses are statements, not theories. At present, we wish to leave room for both approaches, from the known to the unknown, or from the unknown to the known. Neither is characteristic of a theory. (We shall return to this problem in Sec. 4.4.) Now we propose the following provisional *definition* of a theory, as far as its logical structure is concerned:
A theory is a deductively ordered collection of true statements.

Hence, a theory is not just a set of statements, but a qualified collection.[12] This definition is purely logical; it only concerns the formal structure of a theory. Later on, we shall find occasion to expand the definition.

Deduction

A theory is a *deductively ordered* set of statements,[13] meaning that each statement in the set is directly or indirectly connected with each other statement by way of a deductive argument, a deduction. This leads to a *criterion* to decide which statements do or do not belong to a theory: *A statement belongs to a theory if and only if it takes part in the deductive process in the theory.*[14] The above mentioned law of non-contradiction is not the only law concerning the deductive process, characteristic of a theory. A large number of deductive rules or tools of proof are available, such as syllogisms, *modus tollens, modus ponens, argumentum ad absurdum.*[15] Another method is complete induction, applicable wherever numbers are at stake. This method is not universally accepted.[16]

Later on, we shall discuss various kinds of statements in a theory (Secs. 1.5, 4.1). At present, we observe that in most if not all theories, *data* are indispensable for the deductive process. But data are exchangeable. We can replace a datum by its negation. This means that a theory is an *open system*. With our criterion we can decide at any moment which statements belong to the theory, but this can vary from time to time.

Sometimes, a single statement of the type "If...then..." is called a theory. This would not be at variance with our definition, because a set may consist of a single element, and an "If...then..." statement has a rudimentary deductive character. However, a statement like "If *a* then *b*" only becomes meaningful in combination with some other statement, like "It is the case that *a*".

In a technical sense, a theory is a *partially* ordered set, because it is never the case that each pair of statements is connected such that one is deduced from the other one.[17] Each theory consists of a number of independent axioms and data, and a number of theorems, which are derived from the axioms and data. Hence, in a theory two statements may be *directly* connected, if one is deduced from the other, or *indirectly*, either if both are deduced from the same set of axioms and data, or if both are used to deduce a third statement.

The deductive ordering occurring in a theory ought to be non-circular. Circular reasoning is generally rejected, though often applied.

True statements

A theory is a set of statements taken to be *true* within the context of the theory. This is the most intriguing part of our definition. It is a *necessary* part, because of the fact that a false statement allows of any conclusion.

Put otherwise, a theory is required to be *consistent*, *i.e.*, free of contradictions. From a logical point of view, a statement asserted to be false *is* a contradiction.

From a couple of contradictory statements, any statement whatever can be validly inferred.[18] The statement "If p then q" is equivalent to "Either p is false, or q is true". Hence, if we admit both p and its negation, q is always true. Therefore, a false statement cannot be used in a logical process.

On the other hand, very often in a theory statements are used which are known to be false in a wider context. In theories of planetary motion, we now state that the earth is a point, then that it is a perfect sphere, although we know very well that the statements contradict each other, and are both wrong. The "subjunctive" method of using "counterfactuals" is so common, and so fruitful, that we cannot ignore it. Clearly, if we say that we admit only *true* statements in our theory, we do not mean "absolute truth" in whatever sense. We do not even demand that the statements are *believed* to be true — nobody believes the earth to be a point.

Above, we argued that theories are instruments. Instruments are used by people, by logical subjects, in general not by a single person, but by a group of people, who want to use the theory together. These people must decide which statements they want to consider true, for the sake of the discussion. "Let us assume that..." somebody suggests, and the discussion can only proceed if all participants are willing to accept the proposal, if only for the time being.

In other words: statements or propositions are true within a certain context — the context of the theory, and the context of the discussion between the people who use the theory. Outside this context the same statements may be false, or uncertain. But to make the deductive process possible we have to assume that the starting points are true. Then, if we make no logical mistakes, all deduced statements are equally true.

This leads to a most important conclusion: *A theory is never able to prove a statement conclusively*. The truth of any proved statement completely depends on the truth of the axioms and data from which the theory starts. A theory determines the truth of a statement *relative to* the truth of other statements. In how far the latter are true must be decided in a different way. Put otherwise: a theory is an instrument to *propagate* truth, to *transfer* truth, but never to *create* truth.

Three logical relations

We find that a theory functions in three logical relations:

1. In the *logical subject-object relation*, a theory is an instrument between subject and object. A theory is made and used by people, and it is concerned with logical objects, the entities about which people want to theorize. Each statement has a non-logical content, besides a logical form.

2. A theory has a function in an argument, a discussion, a logical debate between people who want to convince each other of the truth or falsity of statements. The participants in the debate must agree on the initial assumptions and the applied methods of proof, because otherwise a discussion would be impossible. We call this intersubjective relation a *logical subject-subject relation*.

3. In using a theory, people are bound to logical rules or logical laws. Hence a theory is indirectly subject to logical laws, it functions in a *logical law-subject relation*. Strictly speaking, it is not the theory which has to obey these laws, but the people who use the theory. Only they can be responsible for any use or misuse of existing or new theories.

In all three relations, logical *subjects* are involved. We cannot and shall not consider theories apart from the people who make them and who use them.

The functions of a theory

From our definition it follows that a theory can never be intended to give a mere *description* of whatever state of affairs. A description can be given with the help of concepts and statements, and usually by a *set* of statements. But a mere description is not an ordered deduction, and does not constitute a proof. A theory has other functions than to provide a description of reality. These functions: to predict, to explain, to solve problems, to systematize our knowledge, will be discussed in the next chapters.

The distinction between a mere statement (or set of statements) and a theory as a deductive scheme may be illustrated by comparing Copernicus with his famous "precursor", Aristarchus of Samos, of whom little is known but his assumption that the earth moves around the sun.[19] This *statement* of heliocentricity should not be confused with Copernicus' heliocentric *theory*, which transcends Aristarchus' statement. Copernicus was aware of this difference, when he wrote: "...let no one suppose that I have gratuitously asserted, with the Pythagoreans, the motion of the earth; strong proof will be found in my exposition of the circles..."[20]

Before starting our discussion of the functions of a theory, let us investigate the logical structure of a theory in relation to the structures of concepts and statements.

1.4. Concepts

Logical reasoning is based on meaningful distinctions and connections. Not only theories, but also concepts and statements have an instrumental, an intermediary function in reasoning. Distinctions and connections are made in our subjective thought, and they concern external, objective affairs. As intermediaries, concepts and statements are formed by men. They are artifacts, like theories, explicit expressions of our thought. As man-made artifacts, they have an objective character, but they are not the primary objects of thought.

Hence theories, statements, and concepts have a logical structure (as instruments of thought), and a non-logical meaning, referring to non-logical states of affairs. In natural thought explicit concepts and statements are not much in need. In theoretical thought they are indispensable.

The first and most fundamental way to make explicit distinctions and connections is to introduce concepts. Classification and conceptualization belong to the first phase of any field of theoretical thought.

Definitions
For a new concept to be introduced in a theory, it must be defined. We introduced a theory as a set of statements. Because a concept is not a statement, it cannot be an element of a theory. But a definition is a statement, and can therefore be an element of a theory. We employ a definition in order to introduce explicitly a new concept into a theory. Usually, however, concepts are tacitly or implicitly introduced. Examples of explicit definitions could be: "This is Venus", or "Venus is a planet", or "Venus is the planet having a period of 224 days". In any theory we find well-defined concepts alongside ill-defined concepts. Each user of a theory can be challenged to clarify his concepts, to distinguish them clearly from other concepts. We know, however, that it is impossible to de-

fine all concepts, because we need concepts to define others. Each theory has a number of "primitive" concepts, which cannot be defined within the theory.

It is sometimes said that a definition is "free". This is not true, if one wants to introduce a new concept into an existing theory. Also definitions are subject to the logical law of excluded contradiction. The definition of a new concept should not contradict the definitions of already accepted concepts in the theory. Any definition should avoid a "contradiction in terms".

Definitions are low-level statements. A set of definitions alone cannot constitute a theory.

Identity

Concepts may have an individual or a universal character. In the first case they establish an identity, in the second case a species or class.

By its identity each thing can be distinguished from every other thing, each event from every other event, or each individual relation from every other individual relation. Since ancient times, Mercury, Venus, Mars, Jupiter, and Saturn were identified as planets, *i.e.*, wandering stars. This means, for instance, that last night and tonight we recognize the same planet to be Mars, even though it has moved on meanwhile. A significant result of Greek astronomy, ascribed to Pythagoras, was the identification of the morningstar and the eveningstar as the same planet, Venus.

The idea of "identity" is subject to the logical law of identity. Each thing is identical with itself, and each event is identical with itself.[21] In the course of a logical argument it is not allowed to change the identity of the things about which one argues. A common fallacy of identity is "equivocation", *i.e.*, to identify what is not identical. An example is the equivocation of Aristotle's "prime mover" with the God of the Bible or of the Koran.[22]

Often, but not always, identified things or events are tagged by a proper name (Venus), or a date (Copernicus' death: 1543).

Classes

Concepts denoting classes have a universal character. For example, we distinguish planets from stars, and both are celestial bodies. Classes and species may refer to things, *e.g.*, minerals, plants, animals; to events; or to relations. For instance, Aristotle distinguished four classes of change or "motion": variation of essence, of quality, of quantity, and of position. Change of position, also called "local motion," was further divided into natural and violent motion. Natural motion was divided into motion towards the centre of the universe, away from the centre, or around the centre. In Aristotle's theory, the first kind of natural motions refers to the class of heavy bodies, the second to the class of light bodies, and the third to the class of celestial bodies (see Secs. 3.1, 3.2).

A system of related classes and subclasses is called a "taxonomic system." It constitutes the barest kind of theory, for if one states that a certain individual belongs to a certain subclass, it can be deduced that it does not belong to other subclasses, and that it belongs to one or more superclasses.

A class concept is well-defined within a theory, if for any thing, event, or relation it is clear whether it belongs to the class or not. Hence, classes also serve to make distinctions. (On the intension and extension of a class, see Sec. 1.5.)

Properties

Class concepts point to things or events of the same kind, and are often indicated by a *noun*, e.g., stars, planets, motions, dogs, lightnings, birthdays. Properties, on the other hand, point to quite different things or events, which have something in common. They are often indicated by an *adjective*, such as red, heavy, light-reflecting.

Properties connect and disconnect classes. They are used to define classes. The property "light-reflecting" connects planets with houses, and distinguishes planets from stars. Properties serve to mark distinctions and similarities.

Aristotle distinguished between "essential" and "accidental" properties of an individual. Its essential properties indicate its nature, its essence, the species to which it belongs. Accidental properties establish the individual's identity. Uniform circular motion around the centre of the universe is an essential property of a planet. But it is accidental that Mars takes about two years, Jupiter about twelve years to complete one period.

Aristotle's distinction between essential and accidental properties served to distinguish classes from properties, which is not an easy matter. For instance, one could define the class of all red things, but this clearly does not constitute anything like a species. Usually, one only speaks of a class if its elements have more properties in common than are needed to define the class. This means that a proper class can be defined in various ways.

Relations

Aristotle's distinction between essential and accidental properties played an important part in medieval discussions. It has concealed the fact that there is another kind of concepts, *i.e.*, relations. A relation is not a property of a single individual, but a property of at least a pair of individuals, or a pair of classes. Aristotelian philosophy had hardly any place for relations. Something is large or small, heavy or light, warm or cold, moist or dry, moving or resting.

Gradually, the Copernicans became aware that these binary "contraries" had better be replaced by relations, such as larger than, heavier than, warmer than.[23] Especially Galileo paid much attention to this matter. He rejected the contrary distinction between heavy bodies and light bodies, by showing all bodies to be more or less heavy. He emphasized that "rest" is not contrary to "motion", but is only a gradation of motion, with zero speed. A falling body, starting from rest, has a continuously increasing speed, varying from zero to the final value.

Aristotle distinguished between quantitative and qualitative properties, and he clearly valued the latter much higher than the former. The Copernican revolution changed this radically. The question of how large something is will sooner be raised in a climate in which the relation "larger than" is more important than the contrary distinction of "large" and "small". More and more the Copernicans became interested in measurable quantities. Measuring means comparing a property of a thing with a similar property of another thing, which is taken as a measure. Measuring standards are, *e.g.*, a yardstick, a thermometer scale, a standard weight.

Operational definitions

The shift from qualitative to quantitative concepts is one of the most striking features of the scientific revolution of the 17th century. This shift has been of consequence for definitions. In Aristotelian physics, definitions concern the essence, the nature of things. These are "conceptual" definitions. "Gravity" is the tendency of heavy bodies to move towards the centre of the universe. In the theories of Galileo and Newton attempts are made to describe gravity with the help of measurable properties like acceleration, mass, and weight. The distinction between heavy and light bodies, so essential for Aristotelian physics, disappeared. Following Archimedes, both Benedetti and Galileo stated that bodies only move upwards if their density is less than that of air.

Definitions determining how a property can be measured are called "operational". Our century has seen a movement (operationism) saying that operational definitions only are fit to determine the meaning of concepts. Its founder, Bridgman, moreover thought that such a definition should unequivocally indicate how the property concerned should be measured.[24] Hence, if there is more than one means to measure a property, we have actually several properties. For instance, if several possibilities to measure the length of a thing are available, we should rather speak of different concepts of length, according to this somewhat extravagant view.

The concept of mass

By way of example, let us consider the concept of mass. In his *Principia*, Newton defined mass or quantity of matter as volume times density.[25] This is clearly an operational definition, not a conceptual one. The concepts of volume and density cannot be multiplied with another. The 19th-century physicist and philosopher, Ernst Mach, the father of logical-empiricism, has taken offence at this definition, because, according to Mach, density can only be defined as the mass of a unit of volume.[26] It is clear enough that Newton did not define density in this way. He did not define density at all, apparently assuming this concept to be sufficiently known, contrary to mass.

"Mass" was a completely new concept, introduced by Newton himself. We find it neither with Galileo, nor with Descartes. For Descartes, the essence of matter was its extension. Being material meant being extended. Hence, quantity of matter was identical with volume. Newton broke away from this view. For him, matter and space were completely different. Therefore, he needed a new definition of "quantity of matter". Keeping silent about the essence of matter, he merely defined how its quantity can be measured: by volume and density conjointly. In the context of his ensuing theory, he argued that mass is a real property of any body, independent of its position. It turned out that mass is not only a measure of the body's quantity of matter, but also of its inertia. In the second law of motion, mass became the proportion of the net force working on a body, and its acceleration. Finally, mass also functioned in the law of gravity. But all this cannot be derived from the operational definition of mass.

Whereas "mass" was a new concept in Newton's time, "density" was definitely not. Moreover, density can be measured independently of any operation related to the division of mass and volume. A century before Newton wrote his *Principia*, theories and experiments about floating, suspending and sinking bodies, introduced by Archimedes, were used by Benedetti and Galileo to define the concept of density. In 1585, Benedetti rejected Aristotle's fundamental distinction between "levity" and "gravity," pointing out that "levity" is caused by the upward force of the medium. A year later, Galileo published *La Bilancetta* (The little balance), describing a hydrostatic balance to determine the average density of bodies or fluids.[27] In 1611 Galileo was involved in a polemic with a number of conservative Aristotelian scholars about the properties of floating bodies, and in 1612 he wrote *Discourse on Bodies on or in Water*.[28] Extended by Torricelli and Pascal, Galileo's views founded hydrostatics and aerostatics.

Hence, for Newton it was obvious to define the new concept of mass using the well-known concepts of density and volume. But his definition may very well be influenced by his atomistic views, according to which he assumed that in denser matter atoms are more densely packed.[29]

The example of "mass" serves to show that an operational definition is not a conceptual one, and to demonstrate that the meaning of a concept can be defined and enriched during the development of a theory. Such a concept is highly theoretical, *i.e.*, its meaning is determined by its theoretical context. We shall return to this matter in the next section.

1.5. Statements and their context

In this section we shall discuss the logical structure of statements in relation to theories. One kind of statements, definitions, we have already met. The most simple statements are those connecting concepts via the copula *is* or equivalents. Aristotelian "predicate logic" is practically confined to statements like "Socrates is a man", "All cows are animals", "Some men are Greek", and "No swans are black." In modern formal logic, the most important operations are negation, conjunction, disjunction, equivalence, and material implication. Together, they constitute "extensional logic", for details of which we refer to textbooks of logic.[30]

Propositions and propositional functions
In formal logic distinction is made between "propositions" or statements, and "propositional functions," in which variables

occur. For instance, "$x^2 + y^2 = z^2$" is a propositional function. It is not a statement of which the truth can be established. It can only be ascribed a truth value if the free variables x, y, and z are bound, for instance by a so-called quantifier. The existential quantifier is symbolized by "$\exists x$" (there is at least one x, such that...), the universal quantifier by "(x)" or by "$\forall x$" (for every x...). Now, $\exists x \exists y \exists z\ (x^2 + y^2 = z^2)$ is a true statement, whereas (x) (y) (z) $(x^2 + y^2 = z^2)$ is false.

The use of variables, first in mathematics, and soon afterwards in physics, is a fruit of the Copernican revolution.

Universal and existential statements
The two quantifiers draw our attention to the important *logical* distinction between universal statements and existential ones. It reflects the *ontological* difference between the law side and the subjective or factual side of reality. Universal statements or law statements refer to laws. Existential or factual statements refer to facts.

Unfortunately, law statements are often called "laws", and factual statements are often called "facts". Newton's "law" of gravity is really a law *statement*. As such it differs from the *law* governing the motion of the planets. Law statements can be true, or approximately true, or false. Laws cannot be true or false.

Only statements occur in a theory, and statements are human inventions. Newton's law of gravity is a statement invented about 1680. But the law determining the motion of the planets around the sun has been valid long before Newton, even long before human beings inhabitated the earth and started theoretical thought. Similarly, the fact that Jupiter has moons was established by Galileo in 1609. Thus the *factual statement* "Jupiter has moons" dates from that year. But the *fact* that Jupiter has moons is presumably much older.

It should be observed that not each "all statement" is a law statement. If I say "All books in my study are catalogued", I do not refer to a law, let alone a universal law of nature. In a logical sense, this statement does not differ from a law statement like Newton's law of gravity. Hence, the logical distinction between universal and existential statements is not identical with the ontological distinction between law and fact.

Meaning, truth and consistency
Whereas a propositional function has no truth value, it should have meaning. In the above example, it is tacitly assumed that x,

y and z are either numbers (*e.g.*, integers), or magnitudes (*e.g.*, the lengths of the sides of a triangle). The expression "$x^2 + y^2 = z^2$" would be meaningless, if x, y and z were persons, for instance.

The logical function of theories is to establish the truth value of statements, relative to the truth of the initial assumptions of the theory, its axioms and data. Moreover, because any statement contains concepts, a theory may also determine the meaning of concepts.

According to the logical-empiricists, it makes no sense to speak of the truth of theoretical statements, and they stress the importance of "meaning analysis". On the other hand, Popper thinks that "meaning" is unimportant, and should give way to the truth of statements.[31] We consider both views one-sided. The truth of statements depends on the meaning of the concepts used in it, and reversely.

Even more confusion arises if one asks whether a theory is true. We have seen that a theory is subject to the logical law of non-contradiction. A single statement can be true or false, but it cannot contradict itself. Only a set of statements, such as a composed statement or a theory, may contain contradictions. But, as we have seen, a theory can only be free of contradictions if all its statements are supposed to be true, and this presupposes that the concepts used have a fairly clear meaning.

Hence, meaning refers to concepts, truth to statements, and consistency to theories. But meaning, truth and consistency are just as much related as concepts, statements, and theories are.

The logical function of a theory, to prove the truth of statements, cannot be seen apart from the reference of statements to laws and facts. For instance, a statement concerning something that in the same context is asserted not to exist is always false, and cannot be used in a theory. We shall return to the problem of truth, meaning and consistency in Chapter 12.

Intension and extension

A class concept involves both an intension (meaning) and an extension (the number of entities belonging to the class).[32] Both depend partly on the theory in which the concept is used. Consider, for example, the concept "planet" as conceived during the Copernican revolution.

Before Copernicus, a planet was defined as a wandering star, a celestial body moving with respect to the fixed stars. Besides Mercury, Venus, Mars, Jupiter and Saturn, both the sun and the moon were recognized as planets. According to Aristotle's cosmological theory, the seven planets move through the zodiac around the centre of the universe, *i.e.,* the earth.

This concept changed radically with the transition to Copernicus' theory. In his heliocentric theory, a planet is a celestial body primarily moving around the sun. Hence, the earth became a planet, and the sun and the moon ceased to be so. Not only the intension of the concept "planet" changed accordingly, but also its extension. The number of planets decreased from seven to six. Moreover, Tycho Brahe's system, in which the sun moves around the earth, and the planets move around the sun, contains only five planets.

The Copernicans introduced the new concept of "planetary system", namely, a central body surrounded by one or more satellites (the term "satellite" was coined by Kepler shortly after the discovery of Jupiter's moons). Copernicus knew two "planetary systems", the solar system and the earth-moon system. Galileo's discovery of Jupiter's moons (1609) was hailed as a significant reinforcement of Copernicanism, because "...now we have not just one planet rotating about another while both run through a great orbit around the sun; our own eyes show us four stars which wander around Jupiter as does the moon around the earth, while all together trace out a grand revolution about the sun in the space of twelve years."[33] It showed that the concept of "planetary system" was not an arbitrary and improbable alternative for the geocentric systems of Aristotle and Ptolemy.

The extension of a concept can be changed without changing its intension, for example, by the discovery of a new planet, like Uranus (1781). Such may be predicted by a theory, as was the case with Neptune (1846).

Theory dependence and autonomy
Hence, the meaning of a concept is partly determined by its theoretical context; it is "theory dependent". But its meaning also has a certain autonomy with respect to the theory.

The extension of the concept of "planet" is partially independent of the three theories mentioned (Aristotle's, Copernicus', and Brahe's). It invariably includes Mercury, Venus, Mars, Jupiter, and Saturn. This is also the case with the concept's intension. Any theoretical definition has to take into account the character of planets as "wandering stars." Only with respect to the sun, the earth, and the moon, do the three theories differ. The difference with respect to intension implies a *partial* shift with respect to the extension.

A similar view can be held with respect to statements. The truth of a statement is only partially determined by the theoretical context. It is possible to have the same statement in different theoretical contexts, with the same truth content. This is highly fortunate, for otherwise we would be unable to compare different theories.

This view expressing partial theory dependence and partial autonomy, both of concepts and of statements, takes a middle course between two more extreme views, logical-empiricism, and historical-relativism.[34] One of the fundamental assumptions of logical-empiricism is the existence of statements and concepts independent of *any* theory. These are so-called observation statements or protocol statements, and observational or empirical concepts. The empiricists strongly believed in observation, in which they found the certainty and trustworthiness of human experience. They assumed the possibility of finding purely empirical protocol statements, independent of any theory. This appears to be in vain. We cannot perform any observation out of the context of our expectations. In particular, for the observations made in scientific observatories or laboratories, elaborate instruments are used, which are developed according to sometimes rather advanced theories.

Criticism of empiricism is exerted by adherents of so-called historical-relativism.[35] Authors like Kuhn, Hanson, and Feyerabend have stressed the theory dependence of concepts and statements to such an extent that they are overbalanced to the idea that the meaning of concepts and the truth of statements are completely determined by the theory in which they function. This means that we cannot even compare two competing theories. It makes no sense to say that the theory of Copernicus is better than that of Ptolemy, because they are "incompatible", they talk about different worlds. The fact that sooner or later the Copernican theory was accepted is not the merit of the theory, but due to more or less accidental historical developments.

In fact, Kuhn, Hanson, and Feyerabend maintain that it is senseless to organize discussions between adherents of competing theories. It may be doubted whether Galileo for instance would have accepted this view. Galileo became famous because of his two great books, the *Dialogue* (1632), and the *Discourses* (1638). These works of fiction describe discussions between three persons, Salviati, Sagredo and Simplicio. Although the opinions of Salviati (the Copernican) and Simplicio (the Aristotelian) differ strongly, they are able to discuss all problems put forward.

Agreement with the view that observations cannot be made apart from *any* theoretical context should not blind us to the fact that observational results may be quite independent of *some* theories. For example, the occasional backward motion of the planets, which played an important part in Copernicus' heliocentric theory, was never disputed by any astronomer, whether he adhered to Copernicanism or not. We shall return to this matter later on.

Arguments

Hence we find that actual theories, statements and concepts are intimately interwoven. This is also the case with their respective structures. Without any doubt, these structures are logically qualified. The main function of theories, statements, and concepts is to mediate between logical subjective thought and its objects.

We consider logic a fundamental mode of human experience, not unlike the numerical, spatial, kinematic, and physical modes, to be discussed in Chapter 3. Because *theories* are characterized by deduction, their structure has a typical kinematic aspect, deduction being the logical movement from one statement to another. Similarly, *statements* have a typical spatial aspect, being dominated by the idea of connection. Statements invariably connect other statements or concepts. Finally, the structure of a *concept* refers to the logical unity and diversity, hence to the numerical mode of experience.

Supposing the numerical, spatial, kinematic and logical aspects to be mutually irreducible, we now understand both why concepts and statements are theory dependent, and why they have nevertheless an irreducible autonomy. In this structural interlacement the meaning of concepts, and the truth of statements, are both relativized and opened up, if they start to function in a theory. Therefore, the use of theories as instruments of thought enriches our experience.

In its turn, a theory has a function in a discussion, an argument.[36] In a discussion, one does not only prove theorems from axioms and data, but the axioms and data themselves are discussed. The force of the arguments is tested, convictions collide, data are weighed. Hence, a discussion (in a logical sense) has a physical foundation, because it is based on a logical interaction between arguments.

This book is mostly concerned with theories, which, however, are incorporated in logical relations. It is impossible to study theories out of the context of logical laws and logical objects, and in particular apart from the logical subjects and their reasoning.[37] It is only in this context that the functioning of theories can be understood. Let us, therefore, proceed with the discussion of two important functions of theories — prediction and explanation. At the same time, we shall have a somewhat more systematic view of the various theories of planetary motion.

2. Explanation and prediction

2.1. Physics and astronomy before Copernicus

In Chapters 2 and 3 we shall be concerned with two important functions of theories, to wit prediction and explanation. Prediction is the first and most obvious aim of any theory. This is a consequence of the deductive character of a theory, *i.e.*, its kinematic foundation, deduction being the logical movement from one statement to another. We shall characterize prediction to be the "kinematic" function of a theory, to be distinguished from its "physical" function, which is to explain. Explanation is tied to a cause-effect relation of some kind.

We have two reasons to discuss explanation and prediction simultaneously. The first reason is the positivist or instrumentalist claim that explanation is structurally the same as prediction.[1] The second reason is that the distinction between prediction and explanation has played an important part in the history of astronomy, both in its pre-Copernican and Copernican stages.

In Chapter 2 we shall discuss the structural difference between prediction and explanation. In Chapter 3 we shall investigate some fundamental modes of explanation, and see that also in this respect the Copernican era was truly revolutionary.

The physics of celestial motion

In ancient and medieval philosophy, distinction was made between physics and astronomy. The first was interpreted in a realistic, the second in an instrumentalistic sense. Physics was the study of the nature or essence of things, and was concerned with the form (Aristotle) or the idea (Plato) which lies at the foundation of each thing or living being. Astronomy was considered part of mathematics; it merely concerned the calculation of planetary motions.

Following Plato,[2] all Greek, Hellenistic and medieval natural philosophers adhered to the view that the perfect celestial bodies can only move uniformly in circular orbits concentric with the universe. On rational grounds Plato argued that celestial bodies move eternally without being disturbed. This would only be possible in circular or rectilinear orbits. Clearly, the celestial bodies do not move on a straight line, which leaves circular motion. Plato's doctrine is conceived to be independent of observation, it is purely theoretical, it is "ideal."

Plato's disciples, Eudoxus and Aristotle, elaborated this view into a system of homocentric spheres.[3] Aristotle referred natural motion to the centre of the universe, towards or from the centre, or around the centre. The linear motions to and from the centre concern sublunary objects, the circular motion around the centre belongs to the heavenly bodies. The transparent spheres carrying the celestial bodies all have the same centre, they are "homocentric." The centre of the universe happens to coincide with the centre of the earth.[4]

The three inequalities

After the establishment of this clear and rational system astronomers were left with the task of reconciling it with the irregularities in the actual planetary motions. This is significantly called "saving the appearances."[4a] The theoretical system was never questioned. In Plato's idealistic philosophy, the phenomena were considered to be deceptive, unreliable and transitory copies of the true and eternal "ideas", according to which the world was made. Aristotle's realism was less extravagant, but the eternal "forms" were also given precedence to transitory phenomena (see Sec. 3.2).

The overwhelming majority of heavenly bodies, the nearly thousand visible fixed stars, satisfy the doctrine of uniform circular motion perfectly well. In 24 hours they turn together around an imaginary axis through the celestial pole. Only the seven wandering stars, including the sun and the moon, show irregularities. In the middle ages, these deviations were discussed under the heading of "inequalities."

The first inequality concerns the seasons. The northern winter is several days shorter than the summer. If the sun would move uniformly in a circle (the ecliptic) homocentric with the earth, winter and summer would be equally long.

The second inequality is the most spectacular, and played the most active part in Copernicus' heliocentric theory. Apart from the daily motion, all planets travel roughly along the ecliptic, and in the same direction as the sun. Except the sun and the moon,

the planets show occasionally backward or "retrograde" motion —
Mercury every 116 days, Venus 584, Mars 780, Jupiter 399, and
Saturn every 378 days.[5] Mercury and Venus, which are never seen
far from the sun, show retrogradation during about half of the
time. The other three planets move backward if and only if they
are seen opposite to the sun, *i.e.*, if the earth is just between the
planet and the sun.

The third group of inequalities consists of all irregularities
which remain if the other two inequalities are accounted for. The
most excessive are those occurring in the motions of Mercury and
the moon.

Ptolemy
The homocentric systems of Eudoxus and Aristotle supplied a
qualitative description of the second inequality, but could not
account for the other two. Moreover, they could not explain the
variations in the apparent magnitudes of the planets. Greek and
Hellenistic scientists have studied many possibilities to describe
all inequalities, using observations and calculations handed down
by Babylonian astronomers since about 1000 B.C. This work of ages
started about 350 B.C., and culminated in the *Syntaxis Mathematica*, "mathematical composition", respectfully called by later
Arab scholars *Al Maegeisty* and now transliterated as the *Almagest*. This book was written by the Alexandrian scientist, Claudius
Ptolemy, about 150 A.D.[6]

In order to arrive at a mathematical description of planetary
motion, including its inequalities, Ptolemy applied three
methods.

The first method concerns the "excenter." The centre of the
ecliptic, the sun's path around the earth, does not coincide with
the centre of the earth. This assumption allowed Ptolemy to take
account of the first inequality, the different duration of the summer and the winter. Excenters were also used in the description of
the other planets' motions.

The second method is the introduction of "deferents" and
"epicycles." In this contraption, a planet does not itself move
uniformly around the earth, but it travels on an auxiliary circle,
the "epicycle." The epicycle's centre moves on the "deferent",
which is the real orbit around the earth. With this device,
invented by Hipparchus about 150 B.C., the second inequality
(retrograde motion) could be accounted for, as well as the
variation of the apparent cross-section of the planets, because of
the varying distance of the planet to the earth.

Ptolemy's third method is the "equant." This device is always used in combination with an excenter. The motion of the planet or the centre of the epicycle is taken to be uniform, not, however, as seen from the earth, but as seen from the equant. This is an imaginary point such that the excenter is exactly halfway between the earth and the equant. (Kepler would later explain why this "bisection of the equant" is necessary.)

These three devices allowed Ptolemy to provide a fairly accurate description of the motion of the planets for ages to come. Later astronomers, both Arabs and Europeans, have corrected and improved Ptolemy's calculations, using new observations. But Ptolemy's methods remained virtually unchanged up till Copernicus' time.

Physics and astronomy
Two points need emphasis. Ptolemy did not speak of spheres, but of circles, which are not homocentric, but heterocentric. And Ptolemy did not attempt to *explain* the inequalities, but to *describe* them. Contrary to physics, mathematical astronomy was interpreted in an instrumentalist sense. The excenters, deferents, epicycles, and equants of Ptolemy's theory were never realistically interpreted; they were never considered to represent the real state of affairs. A faithful disciple of Aristotle, the great 12th-century Arab philosopher Averroës, commented: "The Ptolemaic astronomy is nothing so far as existence is concerned; but it is convenient for computing the non-existent."[7]

Belonging to physics, the homocentric system of Eudoxus and Aristotle was considered by many to be a true and sufficient explanation of the cosmos. Ptolemy's heterocentric system, being a part of astronomy, was merely a useful instrument to make calculations. Most of the time the two theories could peacefully coexist, but occasionally, conflicts between enthusiastic partisans of the two theories could not fail to occur.[8] The former stated certain essential truths about the heavens, it was believed, while the latter better fitted observation.

Besides the descriptive, astronomical *Almagest*, Ptolemy wrote *Hypotheses Planetarum*, a physical and Aristotelian explanation of celestial motion, and *Tetrabiblos*, in which he applied the theory to a practical problem, *i.e.*, astrology. It is the oldest surviving systematic account of astrology, which as a craft was already developed some 400 years earlier. During the middle ages, Aristotle's *On the Heavens* was by far the most influential book on cosmology, however. Ptolemy's *Almagest* was only read by specialists. Shortly before Copernicus, Georg Peurbach and Johann Müller or Regiomontanus were prominent adherents of Ptolemy's methods, whereas Copernicus' contemporary Fracastoro defended a homocentric system consisting of no less than 79 spheres.

Mathematical astronomy was assigned the task of making predictions with a practical purpose. This concerned at first astrological forecasts, later on navigation and the calendar. The annual fixation of Easter, determined by the first full moon in spring, and the adaptation of the length of the year with the help of leap days required knowledge of the motion of the sun and the moon. In the 16th century, the Julian calendar, introduced by Julius Caesar in 45 B.C., was outdated. This fact was the first motive for Copernicus to reform astronomy.[9] The design of the Gregorian calendar (1582), still in use, was not due to Copernicanism, however, although Copernicus' new calculations were used.[10] The needs of navigation became prominent with the voyages of discovery since the close of the 15th century.

The double truth
About 1300, the division of labour between physics and astronomy received a new accent. In the 12th and 13th centuries, translations into Latin of the works of Aristotle, Ptolemy, and others became available in Western Europe, together with Arab comments. The Greek manuscripts, containing many views contradicting Christian doctrines, gave rise to conflicts with the church. For instance, Aristotle taught the cosmos to be eternal and unchangeable, which clearly contradicts the Christian idea of creation. For this reason, during the early middle ages Aristotle was less popular than Plato, who described a kind of creation. In his dialogue *Timaeus*, Plato introduced a divine craftsman, the *Demiurge*, who created the visible world according to eternal "ideas."[11]

In the 13th century, the work of Thomas Aquinas led to a synthesis of official theology with Aristotelian philosophy, including its physics. Since then, philosophy and physics were taught in the faculties of theology of the medieval universities, whereas astronomy as one of the seven *artes liberales* belonged to the substructure of the university, the faculty of arts. The students of liberal arts were free to discuss their views on natural affairs, provided they did not pretend these to be true.

Thus we find people like Jean Buridan, Nicole Oresme, and Nicholas of Cusa, in the 14th and 15th centuries, discussing Aristotle's *On the Heavens*, contemplate the possibility of a daily motion of the earth. (They never considered the annual motion of the earth around the sun.[12] Because this is the most important feature of Copernicus' theory, it is not tenable to consider *e.g.* Oresme a precursor of Copernicus.)

However, as soon as the question arises whether the earth really moves, Buridan writes: "...but I do not say this affirmatively, but I shall ask the lords theologians to teach me how they think that these things happen." And Oresme, Buridan's disciple, who rejected the distinction between celestial and terrestrial matter, and who presented many arguments in favour of the moving earth, in the end rejected its reality, on the authority of Psalm 93, 1: "And yet all people, myself including, believe that the heavens move, and the earth not: Thou hast fixed the earth immovable and firm."[13]

In general, in their comments on Aristotle's works, the medieval scholastics did not question Aristotle's views, but they investigated his *proofs*. Thus, Oresme argued that Aristotle's proof of the immobility of the earth is wanting, but he did not really doubt it. The earthly motion, being contrary to Aristotelian cosmology and biblical texts was considered at most as an astronomical possibility, but never as a physical reality.

The practice of the "double truth" provided the medieval scholars with a certain margin, within which they were free to investigate and discuss anything, if only they ultimately submitted themselves to the authority of the church.[14] But at the close of the middle ages, the authority of the church waned, and with the Renaissance and the Reformation people demanded the right for themselves to decide what is true or false. Hence the practice of double truth became discredited. Copernicus and Kepler rejected it. Galileo became involved in trouble with the church because he refused to adhere to it. Only Descartes still made use of it, to hide his true feelings about Copernicanism, as we shall see later on.

Copernicus

To conclude this historical review, we shall now summarize Copernicus' theory. Although it never became a bestseller,[15] Nicolas Copernicus' book *De Revolutionibus Orbium Coelestium* (On the revolutions of the celestial orbs, 1543) is considered to mark the beginning of modern astronomy. It was preceded by his *Commentariolus* (Short sketch, *c.* 1512), and G. J. Rheticus' *Narratio Prima* (First story, 1540),[16] written by Copernicus' only pupil. A Renaissance scholar, Copernicus looked for inspiration and motivation in antiquity. He wanted to eliminate the distinction between physics and astronomy ascribed to Aristotle. Therefore he returned to Platonic ideas, in order to give a mathematically oriented explanation of the celestial motions.

According to the medieval tradition, there is a sharp dividing line between Platonism and Aristotelianism with respect to the status of mathematics.[17] The Aristotelians attributed to mathematics only a subordinate role. In the 16th century, those who attributed to mathematics a high value for the study of natural affairs called themselves Platonists. Especially in Italy (where Copernicus was a student), they were inspired by Archimedes, another Platonist. His works, published by Tartaglia in 1543, made a strong impression because of their mathematical treatment of physical problems. Copernicus adhered to these Platonic views. "Mathematics (*i.e.*, astronomy) is written for mathematicians", he wrote.[18]

On this account, Copernicus criticized Ptolemy and his disciples for the use of equants, "...an inequality about their centers — a relation which nature abhors."[19] They "...have in the process admitted much which seems to contravene the first principles of regularity of motion...."[20] Besides, Copernicus was motivated by the need for calendar reform, and the search for "...the chief thing, that is, the form of the universe, and the clear symmetry of its parts." He goes on to say: "...if the motion of the wandering stars are referred to the circular motion of the earth, and calculated according to the revolution of each star, not only do the phenomena agree with the result, but also it links together the arrangement of all the stars and spheres, and their sizes, and the very heaven, so that nothing can be moved in any part of it without upsetting the other parts and the whole universe."[21]

Copernicus' system differs from Ptolemy's in four respects.
1. The sun stands still, and the earth moves around the sun, once a year.
2. The planets Mercury, Venus, Mars, Jupiter, and Saturn turn around the sun, just like the earth.
3. The moon turns around the earth. This means that Copernicus' system is truly heterocentric, it has two centres, the sun for the planets, and the earth for the moon. Whereas this was initially considered a flaw of the theory, Galileo's discovery of Jupiter's moons in 1609 was an important reinforcement of the Copernican viewpoint (see Sec. 1.5).
4. The daily motion of the earth about its axis explains the apparent motion of the fixed stars. It makes the earth the *primum mobile*.[22] (In Aristotle's cosmology, the sphere of the fixed stars

was moved by the "unmoved mover," the *primum mobile*. In Copernicus' theory, it is the earth which causes the apparent motion of the fixed stars. In both cases, the prime mover was not considered a *physical* cause, but intended to render motion intelligible.) This is the most obvious novelty of Copernicus' theory. Nevertheless, in Copernicus' two treatises, *Commentariolus* and *Revolutionibus*, the daily motion of the earth is not prominent. It is not his starting point, but rather a consequence of the annual motion of the earth. Still, it is the diurnal motion which became the most controversial feature of Copernicanism.

The first and second points enabled Copernicus to explain the second inequality, the retrograde motion of the planets. For the description of the other inequalities, Copernicus had to take recourse to the same methods as Ptolemy's, with the exception of the detested equant. Copernicus' system is, strictly speaking, not helio*centric*, because the sun is excentric (see Sec. 3.5), but helio*static* — the sun stands still, the earth moves. This is the hall-mark of Copernicanism.

2.2. Prediction

Before Copernicus, the task of astronomy was to describe the motions of celestial bodies. In particular, the main function of an astronomical theory was considered to be the prediction of future positions of the planets, for the sake of the calendar, astrology, or navigation. More generally, the deductive character of a theory allows its use to make predictions. Prediction shares some symmetry properties with local motion.

Symmetry
Predictions concern changeable properties of things or events which are connected in some way. Thus they depend on correlations or coincidences. Literally, "predictions" concern future events, and "coincidences" simultaneous ones, but we shall include "postdictions" or "retrodictions," concerning events in the past. The predictive power of a theory is symmetric with respect to the direction of time.

Also in other respects predictions, based on coincidence or functional coherence, show symmetry. Consider, for example, Galileo's theory of the pendulum.[23] He discovered that pendulums of equal length are isochronous, their periods being independent of the amplitude of the mass of the bob. Next, he found the period to be proportional to the square root of the length of the pendulum. This theory allowed him to predict either the period of a pendulum if its length is known, or to predict its length, if its period is given.

Local motion
The most obvious examples of theories with predictive power concern local motion. The law for the motion of a planet connects its position with the time parameter. Such a theory is Ptolemy's system of deferents, epicycles, excenters, and equants, developed in his *Almagest*. This "calculation machine" did not pretend to *explain* celestial motion. It was only intended to *describe* motions, in order to make *predictions* possible.

Another example is Galileo's theory of ballistic motion.[24] Assuming that the motion of a cannon ball is a combination of two independent movements, Galileo proved its path of motion in a void to be a parabola. The two independent motions include the horizontal one, which is uniform, and the vertical motion, which is once more composed of a uniform motion and a uniformly accelerated one. Galileo succeeded in finding a functional connection between position and time, and between the horizontal and vertical coordinates, such that it became possible to predict the path of the projectile. Again, this theory was not first of all intended to explain, but to predict ballistic motion. Galileo expressly refrained from giving an explanation of the cause of motion.

Empirical generalizations — Kepler's laws
Laws of coincidence or correlations, enabling predictions, are called "empirical generalizations," if they cannot be derived from any accepted theory.[25] They are found by generalization of observed coincidences. Nowadays statistical theories can be helpful in investigating coincidences, but such theories did not exist at the beginning of the 17th century.

For an example, we call attention to Kepler's laws of planetary motion, which had the status of empirical generalizations before Newton incorporated them in his theory of gravity.

Kepler discovered his laws while studying the planet Mars. For a period of about twenty years, Tycho Brahe had observed positions of Mars and other celestial bodies, with a much better accuracy than had ever been achieved before. From 1600 to 1606, Kepler tried to describe the motion of Mars, applying theories of Ptolemy, Copernicus, and Tycho. Ultimately he had to admit his failure. Uniform circular motion being the common foundation of the three theories, Kepler had to conclude that planetary motion is neither circular nor uniform.

Kepler's laws are as follows:

1. The orbit of a planet is an ellipse, with the sun in one of its focal points (1605).[26]

2. Each planet traverses in equal times equal areas, as measured from the sun (1602).[27] This law is now considered an instance of the law of conservation of angular momentum, which was not formulated before the 18th century. Galileo too was aware of this law.[28]

These two laws were published in Kepler's *Astronomia Nova* (New Astronomy, 1609).[29] The third law is found with some difficulty (it is more hidden than published) in Kepler's *Harmonice Mundi* (World harmony, 1619):

3. The relative distance (R) from a planet to the sun is related to its period (T), the ratio R^3/T^2 having the same value for all planets.[30] Later on, a similar law turned out to be valid for the moons of Jupiter, discovered by Galileo (1609), and those of Saturn, discovered by Huygens and Cassini (about 1660). Only the value for R^3/T^2 turned out to be different in the three cases, and therefore depends on a property of the central body. In 1687, Newton proved this property to be the central body's mass (see Secs. 5.1, 6.4).

Kepler's laws, though not constituting a coherent explanatory theory, have predictive power. With the help of the first two laws the position of the planets can be predicted much more accurately than was possible either with Ptolemy's or with Copernicus' system, and also the third law can be used to make predictions.

Kepler's first law is typically Copernican insofar as it constitutes a heliostatic system. The second law contradicts Copernicus' Platonic idea of uniform circular motion, more so than the first law. Even Copernicus contemplated the possibility that the planet's orbit could be elliptical.[31] But the third law is the most typical Copernican achievement. Although it does not follow from Copernicus' theory, it can only be found in a Copernican system, in which the planets, including the earth, are supposed to move around the sun. This will be demonstrated in Sec. 2.5.

2.3. Explanation

To be sure, Kepler did attempt to explain his laws. The full title of his *Astronomia Nova* (1609) reads: "New Astronomy, based on causes, or Celestial Physics, expounded in Commentaries on the motion of the planet Mars." This title expresses the Copernican program to end the division of labour between explanatory physics and descriptive astronomy. Copernicus, Kepler, Galileo, Descartes, and Newton not only wanted to design predictive theories, but first of all they wished to have theories with explanatory power.

An explanation is characterized by giving an *intrinsic, irreversible* and *effective* relation. This is sometimes called *causality*. In an explanation a cause-effect relation is proposed. This *logical* relation is an analogy of physical interaction. Yet it does not necessarily refer to a *physical* relation. We shall discuss several non-physical cause-effect relations.

Intrinsic
By an intrinsic relation we understand something more than a pure coincidence. We speak of a coincidence if we cannot explain it, or if we are sure that there is no explanation. Many events can only be partly explained. For instance, in a traffic accident, there is usually no intrinsic reason why the colliding cars should be simultaneously at the same place. But if at a certain place many accidents occur, people start looking for some intrinsic reason.

In the case of Kepler's third law, the size of the orbit (R) and the period (T) turn out to be related, because for satellites of the same central body, R^3/T^2 has invariably the same value. This leads to the supposition of some intrinsic relation between R and T, a causal relation. Kepler could not find it, and therefore his law remained a correlation for over sixty years.

Also, subsumption under a law or classification does not necessarily constitute an explanation. If we assume that all planets moving around the sun obey Kepler's third law, and we discover a new planet, we can predict that it will also satisfy this law, but we have not explained why.

Irreversible
The irreversibility of an explanation means that cause and effect cannot be exchanged. In the same context, we cannot exchange the *explanans*, the starting point of the explanation, with the *explanandum*, what has to be explained. Making this exchange is a

logical fallacy, called *petitio principii* or begging the question. It means that the *explanandum* is used as an argument in proof. The simplest case is "$a \rightarrow a$," which is logically unobjectionable, but ineffective. It does not convince anybody of the truth of a.[32]

A common version of this fallacy is "labelling" or "name giving."[33] In this case one's reasoning has the form "$a \rightarrow b$", but at close inspection it turns out that a is just another name of b. A famous example is the exchange between Salviati and Simplicio in Galileo's *Dialogue*.[34] Salviati asks "...what it is that moves earthly things downward", and Simplicio answers: "The cause of this effect is well known; everybody is aware that it is gravity", whereupon Salviati replies: "...you ought to say...that everyone knows that it is called 'gravity.'"

Material implication, symbolized by "$a \rightarrow b$", is a logical expression of the irreversibility of the cause-effect relation. However, not every "if...then..." statement constitutes an explanation. Also coincidences, numerical relationships (like Kepler's third law), classifications, and stipulations of the conditions under which a certain law statement is valid, can be given by an "if...then..." statement. An example of the latter is the law of inertia: "If no unbalanced force acts on a body, then it moves uniformly along a straight path."

Effective
The most important requirement for an explanation is that it ought to be *effective*. It should not be able to explain "everything." An explanatory theory which can explain both a certain state of affairs, and its contrary, is useless. An explanation has to distinguish between what has to occur, and what cannot occur. An explanation is imperative, peremptory.

In this respect an explanation differs strongly from a coincidence. Kepler's third law makes R^3/T^2 a constant for all planets, but without further explanation, it could just as well say that R^5/T^2, or $R^3/\log T$, is a constant. An explanation (such as was given by Newton) is only effective if it shows that only R^3/T^2 is a constant, excluding any alternative.

Popper's falsification principle
Popper has stressed the requirement that a theory be effective.[35] He criticizes Marx' theory of history, Freud's psycho-analysis, and Adler's individual psychology, because these theories (according to Popper) can explain everything that happens in their field of research.[36] It is a bit remarkable that Popper

nevertheless has a great admiration for ancient atomic theory, which also can be considered an example of a theory able to explain everything.[37]

Although Popper's examples can be questioned, his idea is correct. It is the basis of so-called *falsificationism*. Popper says that we should never attempt to verify a hypothesis, but to falsify it. We should try to show that the explanation given is unsatisfactory. A theory must be vulnerable, as vulnerable as possible, hence as precise as possible. The chance that a theory is true should be made as small as possible. If it turns out to be impossible to falsify such a theory, then it is not only probably true, but also powerful.[38]

The possibility of falsifying a statement makes it a scientific or empirical one, according to Popper. Falsifiability is a "demarcation criterion", because it demarcates scientific, empirical statements from non-empirical ones.[39] It separates empirical science from pseudoscience (like astrology), non-empirical science (logic and mathematics), and metaphysics. However, non-empirical statements are not necessarily false or meaningless. In this respect, Popper differs from the logical-empiricists, who distinguished empirical (*i.e.*, for them, observationally verifiable), tautological (*i.e.*, logical or mathematical) and meaningless (*i.e.*, metaphysical) statements.

Explanation of phenomena and of single events
Above we distinguished between factual statements and law statements (Sec. 1.5). As far as an explanatory theory is able to deduce both, it can explain laws as well as singular events. We shall pay most attention to the explanation of "phenomena," *i.e.*, repeatable events. In the next section, for example, we shall discuss the phenomenon that retrograde motion of a planet always coincides with opposition of this planet, the earth and the sun. Such a phenomenon has a lawful character. Hence, the explanation of phenomena means the explanation of laws.

The explanation of a single event may proceed by reference to a theory which proves that if certain initial conditions are satisfied, the event necessarily occurs. Next, it may be required to explain why these initial conditions occurred. It is certainly not sufficient to point out that the event to be explained is just an instance of a phenomenon. For example, if somebody asks why tonight's retrogradation of Mars coincides with tonight's opposition of Mars and the sun, he will probably not be satisfied by the mere statement that this is always the case. Actually, this answer could have been given by Ptolemy, because he knew the phenomenon. He was able to predict any single instance of the coincidence, but he could not explain it. Only Copernicus' theory is able to explain the phenomenon, including every single occurrence of it, if it is clear that the initial conditions are satisfied.

2.4. Retrograde motion

The Platonic doctrine of circular uniform motion was taken seriously by Copernicus,[40] and in his theory it plays a much more important part than in Ptolemy's. Copernicus used the principle of circular motion to explain the movement of each planet around the sun, the rotation about its axis, and even gravity. Because the earth is a planet, each planet should be attributed terrestrial properties. Hence Copernicus assumed gravity to be directed to the centre of each planet, instead of to the centre of the universe, as Aristotle taught.[41] The planet's rotational motion about its own axis is, according to Copernicus, a consequence of its spherical shape.[42] Hence, Copernicus used geometrical arguments to explain physical states of affairs.

In the present section we shall discuss retrograde motion, the second inequality, with respect to the superior planets. For the inferior planets, the argument is essentially the same, but it differs in some details.

Retrograde motion and opposition
In Ptolemy's geocentric system it is a coincidence that retrograde motion always occurs if the planet is in opposition. For the superior planets, Mars, Jupiter, and Saturn, this occurs if the earth is between the sun and the planet, and the planet is in its summit during midnight. For the inferior planets, Mercury and Venus, retrograde motion occurs if the planet travels between the earth and the sun, hence it is more difficult to observe.

For the description of this phenomenon, Ptolemy needed a deferent and an epicycle for each planet. By a suitable choice of the relevant parameters (such as the periods of revolution and the relative size of the circles), it is possible to reconstruct the retrograde motion such that it occurs at the right time, *i.e.*, when the planet is in opposition. Simultaneously the planet is closest to the earth, which agrees with the fact that its brightness is a maximum during retrogradation.[43] However, this is not an explanation, and it was never intended to be so. Ptolemy's system would have worked just as well if the backward motion did not coincide with opposition and maximum brightness.

Explanation of the coincidence

Also Copernicus needed deferents, epicycles, and excenters to give an accurate description of celestial motions, but his basic idea differed from Ptolemy's. He assumed that the planets, including the earth, move around the sun with angular velocities decreasing with increasing distance to the sun. Without any further hypothesis, Copernicus could not only explain the phenomenon of retrograde motion, but also why it coincides with opposition and maximum brightness. He observed: "All these phenomena proceed from the same cause, which lies in the motion of the earth."[44]

The apparent retrograde motion occurs when the earth, moving between the sun and the planet, overtakes the planet. At that moment, when the planet is in opposition, it is closest to the earth, and hence its apparent magnitude is a maximum.[45]

The vulnerability of Copernicus' theory

Copernicus' theory differs from Ptolemy's because the coincidence of retrograde motion, opposition and maximum brightness is necessary, it could not have been otherwise. If ever a new planet would be discovered displaying retrogradation without simultaneous opposition, Copernicus' theory would be falsified, which is not the case with Ptolemy's theory. Copernicus' theory is more vulnerable, because it explains something.[46]

Whereas in Ptolemy's theory opposition merely coincides with retrograde motion and *vice versa*, Copernicus' explanation is irreversible. In his theory it makes sense to state that opposition is the cause of the observed retrograde motion, whereas it does not make sense to state that retrograde motion causes opposition. This causal relation has no physical character, however, but a kinematic one. No interaction between opposition and retrograde motion is involved. The real motion of the earth and the planets causes the apparent backward motion, observed from the earth.

This means that Copernicus had realized one of his Platonic ideals. The deceptive, apparent, observed retrograde motion was explained by the cooperation of real, ideal, uniform circular motions.[47]

Prediction of phenomena

If philosophers speak of "prediction," they mean sometimes something different from what we discussed in Sec. 2.2. It is said that a new theory must obtain the prediction of new phenomena, besides the explanation of well-known ones. For instance, Lakatos states that Copernicus' theory was able to predict several new phenomena, and had to be preferred for this reason.[48]

On closer inspection, this is questionable. Copernicus was able to explain a number of well-known phenomena, but he was less fortunate with respect to new ones. He failed to predict — though he could have — the phases of Venus and Mercury, just like Tycho Brahe, whose theory could have predicted these as well.[49]

Copernicus did predict the stellar parallax. This is the same effect as the retrograde motion of the planets, an apparent motion, caused by the real motion of the earth. Copernicus knew that the stellar parallax was not observable, and as a Popperian-before-the-fact, Tycho Brahe rejected the idea of a moving earth for this reason (see Sec. 2.5). According to Popper, if a theory is falsified, one should not take recourse to *ad-hoc* hypotheses in order to save the theory (he calls this a "conventionalist stratagem,"[50] see Sec. 5.2). Copernicus did just that, by assuming that the fixed stars are so far away that the parallax was undetectably small. By this auxiliary hypothesis, he explained the non-occurrence of a phenomenon predicted by his theory. Indeed, only in 1838, nearly 300 years after Copernicus' death, F. W. Bessel measured the stellar parallax applying much better instruments than those available in the 16th and 17th centuries. During the Copernican revolution, the stellar parallax counted more against than in favour of Copernicus' theory.[51]

This kind of "prediction" is more related to explanation, the establishment of a causal relation, than with prediction based on established coincidences. The philosopher Hempel says that explanation has the same structure as prediction.[52] This may be true for the kind of prediction discussed in the present section, but not for predictions merely based on coincidence. If Hempel were right, it would be unexplicable why the Copernicans preferred the Copernican theory above the Ptolemaic one. But we can explain it. It is because Copernicus could explain states of affairs which in the Ptolemaic system were merely coincidences.

2.5. The size of planetary orbits

The transition from the geocentric world view of Aristotle and Ptolemy to Copernicus' heliostatic system is sometimes assumed to be merely a shift of the origin of the coordinate system. This is incorrect. It would be true if we were only concerned with the motions of the sun, the moon, and the earth. In that case it would not matter if these motions were described with respect to the sun or to the earth, or even to the moon, for that matter. For this reason, Copernicanism did not influence the reformation of the calendar, effectuated in 1582.

The most important difference concerns the motion of the planets. In Ptolemy's system, they move around the earth, and in Copernicus' system they move around the sun. From an astronomical point of view this is more important than the question whether the earth moves or not.

Tycho Brahe's system

The Dane Tycho Brahe was the most important astronomer between Copernicus and Kepler, especially because of his excellent observations. He easily recognized the advantages of Copernicus' theory, but he rejected the earth's motion, for various reasons.

1. *Astronomical.* If the earth moves, the stars must show parallax, which was never observed (see Sec. 2.4). Tycho rejected Copernicus' explanation based on the *ad-hoc* assumption that the fixed stars have a very large distance.[53] He also pointed out that the comets do not show parallax, which was only explained by Newton.

2. *Physical.* The motion of the earth should be detectable. Copernicus' defence against this argument was very weak.[54]

3. *Theological.* If literally interpreted, the Bible teaches that the earth stands still (see Secs. 2.1 and 8.3).

However, Tycho accepted the idea that planets move around the sun. In his compromise system, the moon and the sun turn around the earth, and the five planets turn around the sun. This explains retrograde motion in the same way as Copernicus' theory did. (On the difference between the two theories, see Sec. 11.2.)

The power of a theory

Both theories, Copernicus' and Tycho's enable one to make use of the observed size of the retrograde motion (the angular distance of the stationary points) in order to determine the distance of the planets to the sun.[55] This possibility is absent in Ptolemy's theory.

Ptolemy was aware of his inability to determine the size of the planetary orbits. About the three superior planets he assumed that Mars is closest to the earth, followed by Jupiter and Saturn. The argument for this choice is the observed periods of revolution of the deferents. For the sun this is 1 year, for Mars 1.9 years, for Jupiter 12 years, and for Saturn 29 years. This means that Saturn drifts eastward across the stars more slowly than does Jupiter and that more slowly than Mars and that more slowly than the sun. The moon, drifting about 13 degrees per day is closest to us. The reason for this ordering is that in ancient astronomy planets move most unlike the daily motion of the stars the farther they are from the stars. This argument is useless with respect to the inferior planets, because the periods of their deferents are the same as that for the sun, i.e., 1 year. Because Mercury needs 88 days for its epicyclic motion and Venus 224 days, Ptolemy assumed rather arbitrarily that Mercury's position is between Venus and the moon, which period of revolution is 27 days.[56]

This argument provided Ptolemy only with the order of the planets, not with the size of their orbits. If Copernicus would have followed the same argument, he would have placed Mercury closest to the sun, followed by Venus, Earth, Mars, Jupiter, and Saturn. Actually, he found precisely this order, but on the basis of a much stronger argument, which supplied him not only with the *order*, but also with the *relative size* of the orbits. In this respect, the superiority of Copernican astronomy over the Ptolemaic one is obvious.[57]

Copernicus' argument is that the apparent retrograde motion is a projection of the actual motion of the earth around the sun. (In Tycho's system, it is a projection of the motion of the sun around the earth.) This means that the apparent size of the backward motion is inversely proportional to the distance from the planet to the sun.

From the relative distances of the planets to the sun the size of the apparent retrograde path is explained.[58] This is again an example of an explanation, now based on a spatial rather than a physical cause-effect relation, and it is irreversible. From the observed distance of the stationary points, Copernicus could *calculate* the size of the planetary orbits, but not explain them. On the contrary, he explained the size of the observed retrograde motion from the size of something he did not know beforehand, that is, the relative size of the planetary orbits.

Copernicus also calculated the relative orbital speeds of the planets. Ptolemy had tacitly assumed this speed to be the same for all planets, but Copernicus' theory shows it to have a lower value if the distance to the sun is larger. This was beautifully confirmed by Kepler's second law.

Simultaneously, the theory becomes more vulnerable in this

way. For suppose we have means to determine the distance from the planets to the sun (or their orbital speeds) independently; this could be of consequence for Copernicus' theory.[59] Ptolemy's theory would be insensitive to such independent measurements.

Kepler's explanation of the planetary orbital dimensions

Even more than Tycho Brahe, the German astronomer Johannes Kepler was deeply impressed by the power of Copernicus' theory to bring about the dimensions of the planetary system. Already as a student at Tübingen and influenced by his teacher Maestlin, he became a disciple of Copernicus. In 1594, at the age of 23, Kepler was appointed professor of mathematics at Graz. There he developed his theory about the architecture of the planetary system, which he published in his first book, *Mysterium Cosmographicum* (World mystery, 1597).[60] In his theory, Kepler tried to explain two Copernican results, the number of planets (see Sec. 1.5), and the sizes of the planetary orbits.[61]

Kepler arrived at his theory by studying a relatively simple problem, to construct an equilateral triangle's inscribed and circumscribed circles, and to determine the proportion of their radii. Kepler observed that it was close to the proportion of the orbital dimensions of Jupiter and Saturn.[62] Now he constructed within the smallest circle a square, and its inscribed circle, then a pentagon, and so on. This did not lead to anything, because the proportions of the circles obtained did not agree with those of the planetary orbits, and because it yielded an infinity of orbits.

However, Kepler had more success when he applied the same prescription to three-dimensional polyhedra, the same as Plato applied in his theory of the elements (Sec. 3.1). The Greeks had pointed out that only five regular polyhedra exist with, respectively, four, eight, or twenty triangular faces, six square faces, or twelve pentagonal faces.[63] Kepler constructed these five bodies such that the circumscribed sphere of one was the inscribed sphere of the next. This yielded six spheres, one for each planet, and solved his first problem, why the number of planets is six.[64]

Next, he calculated the relative size of the radii of these spheres. This yielded five independent values, three of which agreed fairly well with the values of the planetary orbits calculated by Copernicus. The values relating the smallest two spheres was adjusted by changing the model, and the disagreeing value relating the largest two spheres Kepler waved aside.[65]

Kepler's model can be seen as an example of a spatial explanation. Kepler recognized its limitations. It could not explain the periods of planetary motion. In fact, it was never taken seriously except by himself. However, the publication of his book drew the attention of Tycho Brahe, who was in search of an able assistent to work on the observations made during the past decades. Tycho recognized Kepler's genius, and invited him to Prague, where in 1600 the lifework of Kepler began, the attack on the motion of Mars.

Even after the discovery of the laws which made him famous, Kepler remained faithful to his juvenile work, and he felt greatly relieved to learn that the four new "planets" discovered by Galileo in 1609 were satellites of Jupiter, not of the sun. Even Huygens, when discovering Titan, the first moon of Saturn (1655), thought that this completed the solar system, now containing six moons besides six planets.

In 1619 Kepler published his "harmonic law" connecting the dimensions of the planetary orbits to their periods of revolution (see Sec. 2.2). As we observed, this law could only be discovered by somebody who accepted the Copernican viewpoint. In 1632, Galileo, who completely ignored Kepler's laws, explained the relation between the radii of the orbits and the speeds of the planets from the hypothesis that all planets arrived at their orbits after having fallen from the same point. Newton proved this explanation to be wrong.[66]

Galileo's explanation of the position and motion of the sunspots
As a final example of a non-physical cause-effect relation we point to Galileo's discussion of the motion of the sunspots. Galileo's claim to be the first to have discovered the sunspots is unjustified, but the quality of his observations and his reasoning was unsurpassed during his lifetime. He observed that some sunspots do not change their dimensions during several days, except that their apparent width decreases if the sunspots move from the centre towards the circumference of the solar disc. Based on a careful measurement he demonstrated that this phenomenon could only be explained if the sunspots are situated on the surface of the sun, which is spherical and rotates in about thirty days about its axis. Thus he explained the apparent change of the sunspots by the motion of the sun. In a similar way he explained why the apparent speed of the sunspots decreases during this motion.[67]

The fundamental distinction between prediction and explanation
This concludes our discussion of the distinction between the predictive and explanatory functions of a theory. Prediction is

based on a coincidence of pairs of properties of some kind of events, and is symmetrical in this respect. Explanation is based on a cause-effect relation, which is by its nature asymmetric, and should be effective.

A predictive theory like Ptolemy's needs not be explanatory, and an explanatory theory like Kepler's model based on the five regular polyhedra needs not be predictive. Prediction and explanation are irreducible functions of a theory, because their logical structure is respectively based in two irreducible aspects of human experience, motion and interaction. These we shall discuss in the next chapter.

Not everybody accepts that theories should do more than predict, that theories should have explanatory power. The view that the only function of a theory is to make predictions is called "instrumentalism." Before Copernicus, astronomical theories were interpreted in an instrumentalist sense. Instrumentalism was attacked by Copernicus, and defended by theologians like Osiander and Bellarmine, as we shall see. We like to stress that our view of theories being instruments of thought does not commit us to instrumentalism.

In Sec. 1.5 we observed that theories and arguments differ in a logical sense, because theories have a kinematic foundation in deduction, and arguments have a physical foundation in their force, their weight. In the present chapter we stated that prediction is a kinematic function of theories, explanation a physical one. It means that theories are more "neutral" with respect to predictions than with respect to explanations. In particular instrumentalists stress that a theory can be used to make predictions, even if its basic axioms are false. In this way, the Ptolemaic system was accepted by the Aristotelians. But this view can hardly be accepted with respect to explanation. Everybody feels that a theoretical explanation cannot be effective if the basic axioms of the theory concerned are taken to be false.

It will be clear that we used the words "kinematical" and "physical" in an analogical sense. We considered deduction as *logical* motion, and arguments as being concerned with *logical* force. Nevertheless, it may be useful now to study the *original*, non-logical meaning of motion and interaction, as we shall do in the next chapter.

3. Principles of explanation

3.1. The harmony of the spheres

There are some reasons to interrupt our discussion of the functions of a theory. First, we have already found occasion to allude to the idea of "fundamental modes of human experience" (Secs. 1.2, 1.5, 2.5), which also play a part in chapters to come. Secondly, we shall see that the Copernican revolution was instrumental in the development of two of these irreducible aspects of experience, to wit the aspect of pure, *kinematic motion*, and the aspect of *physical interaction*. Before the Copernican era, especially the *numerical* and *spatial* aspects were crucial, both in ancient and medieval philosophy. We shall discuss these first (Secs. 3.1, 3.2). The modes of experience are simultaneously modes of being, of temporal relations, of thought, and modes of explanation. We shall especially pay attention to the latter, the fundamental and irreducible principles of explanation. But we shall not treat them systematically, but historically, *i.e.*, in the historical context of the Copernican revolution.[1]

Copernicus, Galileo, and especially Descartes, propagated the explanation of motion by motion (Secs. 3.3, 3.4). In addition, Kepler and Newton explained changes of motion by a force (Secs. 3.5, 3.6). This development was stimulated by the leading theme of Copernicanism, the moving earth.

Pythagoras and the rational numbers

In the 6th century B. C., Pythagoras and his school were looking for a rational description of natural phenomena, in terms of ratios of integral numbers. "Rational" means both "reasonable" and "proportional." Rational numbers are intelligible and proportional numbers.

For instance, the Pythagoreans discovered that the musical tones which sound harmoniously together, stand to each other as integral numbers. These tones were not related to frequencies or wavelengths, as they are considered now, but to the linear dimensions of musical instruments. Two strings of unequal length producing different tones differ by an octave if the lengths of the string are as 1:2, by a fifth if the proportion is 2:3, whereas a fourth corresponds to the ratio 3:4.

The Pythagoreans attached much significance to "complete" numbers such as 10, being the sum of 1, 2, 3, and 4. These four numbers were supposed to determine a point, a line, a plane, and a body, respectively. Accordingly, the Pythagoreans assumed ten celestial bodies or celestial spheres. The earth, the moon, the sun and the five planets were taken to be spherical and to turn around a central fire. In order to arrive at the number 10, some postulated a counter-earth, moving on the other side of the central fire, and hence never being observed.[2] The central fire is never seen because the earth is only inhabitable at the side turned away from the fire.

The motion of the seven planets stand to each other as the notes of the octave, displaying a harmony, the harmony of the spheres. Even Kepler was inspired by this view. In his *Harmonice Mundi* (World harmony, 1619), he employed musical notation to describe the excentric motion of the planets.[3]

Plato and space
Plato based a theory of matter on the Pythagorean discovery of the existence of exactly five regular polyhedra (see Sec. 2.5). The four elements, earth, water, air and fire, introduced by Empedocles, respectively correspond with the cube, the icosahedron, the octahedron, and the tetrahedron. The dodecahedron characterizes the fifth element, *quintessence* or ether, from which the heavenly bodies are made. The nature of this correspondence is quite vague.[4]

Although influenced by the Pythagorean tradition, Plato made an important shift. Whereas the Pythagoreans stressed *numerical* proportions, Plato emphasized *spatial* relations. For the Copernicans, the Pythagorean-Platonic tradition meant a source of inspiration to build a mathematical physics.

The shift from the numerical to the spatial mode of explanation was caused by a crisis in the Pythagorean movement, which led to its disbandment.[5] This crisis occurred shortly after the discovery of the famous Pythagorean theorem, implying in a square with side 1, the square of a diagonal equals 2. The Pythagoreans succeeded in proving that the length of this diagonal cannot be expressed by a rational number. The length of the diagonal is thus not rational, it is irrational, unreasonable, unintelligible.

The starting point of the Pythagorean school, to explain everything with the help of rational proportions, was shipwrecked on a simple spatial problem. The Pythagoreans could not overcome the irreducibility of spatial relations to purely numerical ones. As a result, mathematicians inspired by Plato, like Euclid and Archimedes, directed themselves to the development of geometry.

Aristotle on natural motion

Before Galileo's, the most important treatise of motion was Aristotle's *Physics*.[6] It can only be understood in its relation to his cosmology, and the theory of the elements, displayed in Aristotle's *On the Heavens*. Just like Plato, Aristotle put the earth in the centre of the universe.

If the cosmos were in equilibrium, it would consist of a set of perfect concentric spheres. At the centre is the sphere of the heavy element earth, surrounded by the less heavy element water, the light element air, and lightest of all, fire. In or near the sphere of fire we observe lightning, comets, meteors, and *aurora borealis*. It occupies the periphery of the sublunary spheres rather than their centre, as the Pythagoreans assumed. The sphere of fire is delimited by the lunar sphere, the lower boundary of the heavenly space in which we find the celestial bodies.

According to Aristotle, the lunar sphere constitutes a sharp division between the celestial and terrestrial realms. The celestial space is ordered, the sublunary space is disordered. In the heavens, everything is perfect — the spherical shape of the celestial bodies, their unchangeability and incorruptibility, their circular uniform motion. On the other side, the sublunary sphere is imperfect. The separation of earth, water, air and fire into concentric spheres is disturbed, and the four elements are mixed. Here one finds not only natural motion, but also unnatural, violent motions. In the present section, we restrict ourselves to "local motion."

Natural motion in the sublunary spheres is vertical and rectilinear. It is directed toward the centre for heavy bodies, and toward the periphery for light bodies. Even this natural motion is caused by an unnatural, artificial state, deviating from the ideal state of equilibrium. Natural motion means motion to the natural place, and can only arise if the moving body is away from its natural place. This shows that also for Aristotle spatial relations have a high priority as principles of explanation. Natural motion is explained by spatial arguments.

The natural motion of the heavens, too, is not without cause. It is caused by the "unmoved mover," the god who, at the uppermost periphery of the cosmos, is nothing but pure thought, thinking about itself — thought returning to itself. The rotation of the heavens is caused by the love of this god, by striving to become god-like, perfect. Circular motion is returning in itself. Ultimately, all motion is caused by the prime mover, and the centre of the cosmos is unmoving.

Aristotle's cosmos is an organized whole, in which everything has its natural place. It is, however, not a living organism. Aristotle did not adhere to astrology, which was not influential in Athens of his time. However, when during ages to come astrology and alchemy came to the forefront, their leading idea of the microcosmos-macrocosmos correspondence easily fitted into Aristotle's cosmology.

Violent motion and the impetus theory

Aristotle distinguished natural motion from violent, artificial motion, motion influenced by a force.[7] Natural motion is motion according to the nature of a thing, whether it is heavy or light. A heavy body falls downward, because it is heavy. This is an internal cause. For a natural motion no external cause is needed, as it is for any unnatural motion.

Violent motion is proportional to the driving force, and is inversely proportional to the resistance it receives, at least, if the force is large enough to overcome the resistance.[8] This proportionality should not be taken in a strictly numerical sense. Aristotle apparently did not aim at a mathematical approach to physics.[9]

The relation between natural and violent motion is rest. Natural motion is the actualization of some potential (see Sec. 3.2), and naturally ends when the body has achieved its end, its natural place. Violent motion is contrary to the nature of a thing. Therefore, by force of its nature, everything resists violent motion. For Aristotle, rest is ontologically different from motion, both natural and violent. In fact, it is ontologically superior to both.

During the middle ages, Aristotle's theory of local motion caused much discussion. In particular, it is by no means clear what kind of force causes the motion of an arrow. Apparently, it is the force of the bended bow. But this force ceases to work as soon as the arrow has left the bow, whereas the motion does not cease, and Aristotle rejected any kind of action except action by contact.[10]

Several solutions to this problem were considered. The most interesting is the impetus theory, in the 14th century developed by Buridan and others at Paris. According to this theory, motion can be caused by an internal or an external motor. The external motor is the force, the internal motor is called the "impetus." The bow does not only supply an external force to the arrow, but also an internal impetus. During the motion, the impetus decreases until it is exhausted, and the motion ceases. In the theories of Aristotle and his disciples it was unimaginable that a body would partake in a natural and a violent motion simultaneously. An arrow shot obliquely would move rectilinearly until its impetus is exhausted. Only then it would begin to fall.[11] The observed *curved* path of motion is contradicted by the theory, and hence deceptive.

Impetus is proportional to the quantity of matter in the body and its motion. In the 17th century this was transformed into mass and velocity, but these magnitudes were not defined yet in the 14th century. Nevertheless, the impetus can be recognized as a predecessor of the modern concept of linear momentum. However, impetus was considered as the cause of motion, whereas the later momentum is merely a measure of motion, quantity of motion. The transformation of impetus into momentum has been a laborious process, and is a fruit of the Copernican revolution.

Although the medieval impetus theorists assumed that a falling body is increasing its impetus, and they also studied uniformly accelerated motion, before the 16th century the two were never related.[12] But Buridan arrived at the important insight that the speed of a falling body at any instant depends on the path traversed since the start of the motion, rather than on the distance to the body's natural place.[13]

During the middle ages, projectile motion was never considered a proof against the validity of Aristotle's views. Even the impetus theorists tried to solve the problem within the context of Aristotle's physics. Only in the 17th century did it lead to the emergence of a new theory of motion. This was only possible after the distinction between celestial and terrestrial physics was destroyed, *i.e.*, after the introduction of a new cosmology — the Copernican one. But before we enter into this matter, we shall first consider Aristotle's theory of change more closely.

3.2. Explanation of change in Aristotelian philosophy

In Copernicanism, a new principle of explanation emerged, *i.e.*, motion. In order to make clear the revolutionary character of this novelty, we should emphasize that in ancient and medieval philosophy motion was never a *principle* of explanation, but always the *result* of an explanation. Like any kind of change, local motion had to be explained, and could not be used as *explanans*.

The Aristotelian scheme of explanation was one of the answers given to a question put by Parmenides (*c.* 500 B. C.). On logical grounds, Parmenides proved change to be unintelligible, and thus non-existing. The paradoxes of motion, put forward by Parmenides' pupil Zeno, are well-known exercises to challenge the reality of change.

Form and matter

According to Aristotle, the explanation of change must start from two unchangeable principles, eternal form and eternal matter. Taken individually, both form and matter are universal. The "forms" are very similar to Plato's "ideas" (see Sec. 12.1). The difference between Plato and Aristotle arises with respect to the relation between "forms" or "ideas" and observable things. In Plato, the observable things are unreliable copies of the ideas. In Aristotle, form and matter are united in every individual, in every "substance." This means that Aristotle is more realistic than Plato, and that observations are valued higher. Plato's ideas should primarily be understood in a mathematical sense. The ideal triangle, a subject matter of geometry, is only approximately realized in triangular bodies. Aristotle was primarily inspired by his biological studies. His forms refer to species of animals or plants. Unlike mathematical ideals, these can only be studied by careful observation.

Any individual is a combination of form and matter, is formed matter, is substance. Forms and matter are eternal and unchangeable. Only substances can change.[14] This view prevented Aristotle and his followers from appreciating that also relations and even motion itself might be variable.[15]

Strictly unformed matter, *materia prima* or first matter, is characterized as "absence of form" and therefore does not exist in any concrete sense.[16] But matter can also be conceived in a less absolute sense. When a sculptor creates a sculpture, its design is changed as little as the marble, the material used. Only the concrete piece of marble changes, because the sculptor adds a new form to its matter. This form existed beforehand in the imagination of the artist, and the perfect form does not change. Changing a substance is the process (*kinesis*) by which it attains a new form.

Potentiality and actuality

In order to explain such a process, it is not sufficient to have insight into the form and matter concerned. The matter involved must have the potential to attain the appropriate form.[17] It is possible to make a sculpture out of a piece of marble, but not of sand or water. Marble has the potential to become a statue, which water has not. This is even more clear in living nature. A chicken-egg has the potential to become a chicken, and never becomes a duck or a horse.

Besides form, matter and potential, a fourth principle of explanation is needed, the efficient cause, the actualization of the potential. The chicken-egg must be hatched, the marble must be worked on. Only if these four "causes" are found is the explanation complete, according to Aristotle.

Instead of "potential" and "actual," Aristotle often speaks of the "final cause" or destiny, and the "efficient cause." In this case, the process from potentiality to actuality is treated separately.[18] It is not always possible to distinguish the four causes, sometimes the final cause or end being identical with the form to be achieved.

The Aristotelian doctrine that every motion needs a cause is based on common sense, and turned out to be a great hindrance to the emerging new science of motion, based on the counter-intuitive idea of inertia.

Four kinds of change

Aristotle distinguished four kinds of change in individuals, in decreasing order of importance:[19]

1. Change of essence or nature, generation and corruption, coming into being and passing away, *e.g.*, the birth of a caterpillar, or the death of a butterfly.

2. Change of quality, alteration, *e.g.*, the change of a caterpillar into a butterfly.

3. Change of quantity, increase or decrease, *e.g.*, the growth of a caterpillar.

4. Change of position, *e.g.*, the local motion of our caterpillar on a tree.

The first kind of change is treated in Aristotle's *On Generation and Corruption*, the other three form the subject matter of his *Physics*. Although the least important, the fourth motion takes precedence over the other three, no change being possible without local motion.[20]

Every motion has a beginning (its matter) and an end (its

form). The first kind of motion has "contradictories" for its *termini*, *x* and non-*x*. The other three have "contraries" as *termini*. A subject *x* acquires the property *a*, which it lacks before the change occurs. Also local motion is limited by its *termini*. Therefore, Aristotle rejects the infinity of the cosmos. If the cosmos were infinite, the elements could move infinitely far away, *i.e.*, without an end. There is only one kind of motion having neither beginning nor end — uniform circular motion.[21] It lacks contraries, it lacks potential existence. Celestial bodies, moving uniformly in circles, are completely actual, hence unalterable, eternal, incorruptible, ungenerable, unengendered, impassive. It may be stressed that these cosmic properties are found as a *logical* consequence of Aristotle's theory of change. Even in this case, local motion is the basis of the actuality of the bodies concerned.

In Sec. 3.1 we observed that for a natural motion no external cause is needed. The same applies to the other three kinds of change. We still distinguish between natural and unnatural death — the latter needs explanation. The alteration of an acorn into an oak is a natural process, and does not need an external cause. But the change of an oak into a pile of boards is an unnatural process, in need of an external cause. The free fall of a heavy body can be prohibited, if it is sustained. Similarly, natural processes like the growth of a plant can be prohibited, for instance, because of lack of water. But as soon as such external prohibitions are removed, the process will occur according to its nature, without any external cause.

The four elements

Besides four causes and four types of change, Aristotle distinguished four terrestrial elements. Plato related the elements to the regular polyhedra, but this could not serve Aristotle's theory of change.

Generation and corruption always involves a mixing of elements. The celestial bodies are made of a single element, because they cannot be generated or corrupted. Aristotle related the terrestrial elements to *termini* of change, *i.e.*, pairs of contrary properties — warm and cold, dry and moist, up and down. Earth is dry and cold, water moist and cold, air hot and moist, fire hot and dry. Earth and water are heavy, and by their nature move downward. Fire and air are light, moving upwards. The upward and downward motions are opposite, hence point to imperfection, and to the existence of at least two elements, heavy earth and light fire.[22] The contrary qualities of heavy and light were never related to density. Only neo-Platonic scholars studied density as a quantitative property (see Sec. 1.4).

The four-element theory, if severed from the distinction between gravity and levity, is consistent both with Aristotelian and Copernican views. Up till the 19th century it was the basis of medical and psychological theories. Galileo connected the elements with the senses.[23] As to the theory of matter, the four-element theory was only defeated in the 18th century, especially by Lavoisier.

The theory of the elements enabled Aristotle to criticize the older atomistic views. As a reply to Parmenides' denial of variability, the atomists accepted only local motion as possible change. In various ways, from the time of Galileo most Copernicans were atomists. They considered atomism not in the first place as a theory of the structure of matter, but as the ontological foundation of their world view, in which local motion does not play a subordinate, but rather a leading part.

Aristotle's theory of change is one of the most beautiful intellectual achievements ever made. Apart from being false, it has only two flaws: projectile motion (see Sec. 3.1), and the uncertain status of light (see Sec. 11.3). It is not only completely logical, it is also in harmony with common sense.

Zeno's paradoxes

Parmenides' disciple, Zeno, c. 450 B. C., tried to prove the untenability of motion by means of the well-known paradoxes of motion.[24] One of them is the statement that Achilles, a legendary athlete, would never be able to overtake a tortoise. For if Achilles would have covered the distance which at first separated him from the tortoise, the animal would have moved on.

Concerning the paradoxes of local motion, several views are conceivable.

First, it may be admitted that the arguments are correct, and that motion is illusory. This view was shared by Parmenides, Zeno, and Plato. The observable world to which motion belongs is deceptive, is merely an appearance, a shadow of the real, unchangeable ideas.[25]

Secondly, one may attempt to show that the arguments are incorrect, and that motion is no illusion. This was the aim of Aristotle's theory of change.

Thirdly, Zeno's arguments may be accepted, but it is nevertheless maintained that motion is real. This means the recognition that motion cannot be explained, at least not starting from the categories Zeno applied. In his arguments we find only numbers and spatial distances as elements. Numerical and spatial concepts and their relations are supposed to be the only acceptable principles of explanation. Zeno succeeds in proving that these principles are not sufficient to explain motion.

Shortly before Zeno, the Pythagoreans proved that spatial relations cannot be explained by numerical relations (see Sec. 3.1). Zeno's paradoxes can be interpreted to demonstrate that motion cannot be explained by numerical and spatial relations. The third way to consider Zeno's paradoxes means to accept motion as an *unexplained principle of explanation*. This became the route of the Copernicans. Whereas the Pythagoreans were confronted with the irreducibility of the spatial mode of explanation, Zeno stumbled on the irreducibility of the kinematic mode of experience. But only the Copernicans accepted the challenge head-on.[26]

3.3. Galileo on motion

In Copernicanism, the study of motion plays a central part. The characteristic trait of the Copernican system is not its heliocentrism, but the assumption that the earth is moving. The principal objection against Copernicanism concerned this doctrine. In order to avoid it, Tycho Brahe proposed his compromise system (Sec. 2.5). As an astronomer, he recognized the advantages of Copernicus' theory, and in his own system, the sun has a much more prominent position than in Ptolemy's. But the physical and theological objections against the earth's motion Tycho considered insurmountable.

On the other hand, the genuine Copernicans felt inspired by the difficulties engendered by the earth's motion. One of the most important Copernicans was Galileo Galilei.[27] Between 1609, when he made his most provocative astronomical discoveries, and 1633, when he was convicted by the Inquisition, he was the main agitator in favour of Copernicanism. After this episode, he published his ideas on the theory of motion in his *Discorsi* (1638) which avoided mentioning Copernican views. Most work on his theory was done at Padua, where he was a university professor from 1592 to 1610.

The attack on Aristotle's cosmology
Above we stressed that Aristotle's theory of motion is strongly attached to his cosmology. Therefore, any new theory of motion could only have a chance after Aristotle's cosmology was abol-

ished. Meanwhile, Copernicus' system could only be considered an interesting mathematical exercise. Kepler's attempt to replace the Aristotelian system by a Platonic one (see Sec. 2.5) was abortive, because Plato shared Aristotle's distinction between celestial and terrestrial physics. The absolute split between heaven and earth constitutes the heart of Aristotelian cosmology — so much so that Galileo found it necessary to devote the entire First Day of his *Dialogue* (1632) to the demolition of this distinction.

Galileo's main argument is to show, on the basis of observations, that the celestial bodies are not perfect, not incorruptible, and not unalterable. He extensively discusses the mountains on the moon, in order to show that the moon is not perfectly spherical.[28] He shows that the sunspots are continually generated and corrupted.[29] He points to Tycho's observations showing that the new generable and corruptible stars (*novae*) are celestial objects, not sublunary ones.[30]

On the other hand, he argues that circular motion pertains to terrestrial as well as to celestial objects. This concerns first of all Galileo's conception of inertial motion (see below). In his discussion of magnetism, Galileo lets Sagredo state that a magnet exerts both circular and linear motions, and should therefore, according to Aristotle, be composed of celestial and terrestrial matter.[31]

Despite its prohibition in 1633, Galileo's *Dialogue* terminated Aristotle's cosmology. After him, Descartes, Huygens, Newton, and other Copernicans no longer bothered to refute it.

Motion: a state
From the Copernican viewpoint, motion is a state of a system, existing in itself, *sui generis*, not in need of explanation.[32] A body, moving or at rest, is completely unaffected by which of these two states it is in, and being in one or the other in no way changes it.[33]

Therefore, rest and motion are not contraries, as was taught by Aristotle. Both are states of motion.[34] Although this was only gradually realized, it implies that the attribution of the state of rest or motion to a given body is only possible in relation to another one. Moreover, one motion is unaffected by another, two motions never interfere with each other.

One of the first clear published expressions of the principle of inertia is Galileo's (1613): "And therefore, all external impediments removed, a heavy body on a spherical surface concentric with the earth will be indifferent to rest and to movements toward any part of the horizon. And it will maintain itself in that state in which it has once been placed; that is, if placed in a state of rest, it will conserve that; and if placed in movement toward the west (for example), it will maintain itself in that movement."[35]

In this respect, Copernicanism differed profoundly from Aristotelianism, in which motion is a *process* from potentiality to actuality. In Aristotelian philosophy, motion must be explained, it can never serve as an explanation. The Copernican revolution witnessed the slow and gradual transition from motion as a process to motion as a state. The first hesitating steps were taken by Copernicus himself, who contended that the earth's daily rotation is not in need of any explanation besides the spherical shape of the earth. The natural motion of a sphere is rotation.[36] This shows that Copernicus still accepted a formal cause. The final steps were made by Newton, who associated the idea of inertia with the idea of mass, and by Leibniz, who rigorously stated the relativity of motion.

Explanation of motion by motion
"Explanation of motion by motion" means the adoption of one or two kinds of motion as irreducible principles in order to explain other kinds of motion. It is not necessary to explain the primary or natural motions themselves. The principle of inertia could be formulated as: a body on which no external unbalanced force is acting, *moves because it moves.*

In Galileo's *Discorsi* we find two kinds of primary or natural motion. Both occur without external cause, and are idealized states. First, uniform circular motion is primary, not in need of explanation. It concerns planets turning around the sun, and the earth rotating about its axis, as well as a terrestrial body moving without friction on a horizontal plane, for instance, a ball on a smooth surface, without air resistance. Galileo describes the horizontal motion sometimes such as to be approximately rectilinear, but in principle it is circular.[37] (On Galileo's views on circular motion, see also Sec. 9.4.)

The second natural motion is free fall in a vacuum.[38] In this case, acceleration instead of velocity is a constant. Again, no external cause is needed. Gravity, the source of the motion, is an intrinsic property of the falling body,[39] but not its cause, for the motion of fall is independent of the weight of the body. Galileo found this shortly before 1610.

Galileo's two kinds of natural motion are not contrary to each other. First, one kind can change into the other. A ball uniformly accelerated on an inclined plane may continue its motion at constant speed on a horizontal plane. Secondly, they have a common source, for gravity is the source of all motion. When discussing the inertial motion of a ball on a horizontal plane, Galileo lets the motion start from an inclined plane.[40] The uniform motion of the planets he explains with the help of a Platonic myth on the fall of the planets from a common point toward their present orbits.[41] Gravity is even considered a measure of motion.[42] Thirdly, two natural motions can be composed, a body is able to perform two motions simultaneously.[43] This applies to a ball rolling down an inclined plane,[44] or to the earth combining its daily and annual motions. It also applies to the combination of a horizontal uniform motion and a vertical accelerated motion. This enabled Galileo to explain the path of a cannon ball, i.e., the time-honoured problem of projectile motion. He emphasized that his theory is able to explain why the cannon reaches farthest if the ball is fired at an angle of 45°. This fact was empirically known for quite a long time, but never explained.[45]

Galileo used motion to explain familiar phenomena, such as sound. Following Benedetti, he explained sound as motion caused by the periodic motion of a string, "...the waves which are produced by the vibrations of a sonorous body, which spread through the air, bringing to the tympanum of the ear a stimulus which the mind translates into sound."[46]

Galileo's "instrumentalism"
Galileo's conscious decision to abstain from a *physical* explanation of free fall is sometimes described as positivism or instrumentalism. Allegedly, Galileo restricted himself to a description of motion. This interpretation betrays the view that any explanation needs be a physical one. In Sec. 2.3 we stressed that an explanation in a logical sense has a *causal* character, but this is not necessarily a physical one.

To consider Galileo an instrumentalist is highly improbable, considering his incessant struggle against an instrumentalist interpretation of Copernicanism.[47] If he had been an instrumentalist, he would never have had a conflict with the Inquisition (see Sec. 9.2).

Rather, we should assume that his decision to exclude physical elements from his theory of motion was prompted by his striving to make the most of motion as a principle of explanation. Therefore he rejected the Aristotelian explanation of free fall by gravity.[48] His axiom saying that free fall implies a constant acceleration equal for all bodies enabled him to explain a large variety of phenomena.

The law of fall

Before Galileo, Benedetti had argued that in a vacuum all bodies having the same density would fall with the same speed.[49] Initially Galileo adhered to Benedetti's view that this speed is determined by the density of the falling body. In general, he supposed the speed of fall to be proportional to the difference between the densities of the body and the medium in which it falls. By this law he tried to account for the upward force exerted by the medium, and to disprove the distinction between light and heavy bodies, essential in Aristotle's physics.

Later on Galileo realized that the motion of a body in a vacuum does not depend on its density.[50] Experiments with balls of various density, rolling down an inclined plane, made him see that any influence of the density on motion must be ascribed to the medium.[51]

The same experiments led him to the insight that the velocity at any instant is determined by the time elapsed since the start of the motion, not by its distance to the end point. This refutes Aristotle's view of fall as a motion toward the body's natural place.[52] Initially, Galileo made a mistake by connecting the velocity with the *path* covered since the start of the movement.[53] Shortly before 1610 he arrived at the insight that the increase of velocity is proportional to the *time* passed since the start.[54] This became his definitive law of fall. For all bodies moving in a vacuum, whether vertically or along an inclined plane, the velocity increases proportionally to time, the proportionality constant depending on the angle of inclination.[55]

With this law, Galileo was able to explain an experiment done about 1604. He established that the distances, covered in equal times by a ball on an inclined plane, are in the same proportion as the odd numbers.[56] This means that the path covered since the start is proportional to the square of the time passed.

As a result, after Galileo *time* became the most important parameter in mechanics. It symbolizes the shift from space (the covered path) to motion (the kinematic time) as the main principle of explanation. Therefore, after Galileo the problem of the measurement of time became very urgent, for navigation as well as for mechanics and astronomy. It also created the problem of the nature of time itself (see Sec. 3.6).

The earth's motion and the tides

In his *Dialogue*, Galileo takes pains to make the double motion of the earth acceptable. He does so in two ways. First, he disproves the advanced arguments against the moving earth.[57] Why don't we feel a constant eastern wind? Because the air turns around with the earth. Why does a falling stone arrive at the foot of a tower, and not slightly to the west? Because the stone partakes in the motion of the earth. Here, Galileo applies the relativity of motion.[58] Thus the unobservability of the earth's motion is explained by the motion of everything that moves together with the earth. Galileo's arguments refute the objections against the earth's motion, but do not prove it.[59]

Therefore, Galileo next looks for positive evidence, for phenomena only explainable by the motion of the earth. He finds them in the retrograde motion of the planets (Copernicus' argument), the observed motion of the sunspots, and eventually in the not yet observed stellar parallax. Galileo thought that also the tides are such a phenomenon. He was so much convinced of this that he wanted to give his *Dialogue on the two chief world systems* (1632) the title *Dialogue on the tides*. This was refused by the papal censor, who agreed to an "impartial" discussion of the structure of the cosmos, but who would not allow the suggestion that the earthly motion could be proved. Galileo discusses his tidal theory in the Fourth Day of the *Dialogue*.[60]

He rejects Kepler's view that the tides are caused by the moon.[61] According to Galileo the double motion of the earth, daily and annual, causes accelerations and decelerations, leading to a periodic motion of water in its basin. He stresses that the details of the tidal motion depend on the shape of the basin, by which he tries to meet the objection that his theory implies a period of one day, whereas the actual frequency is twice a day.[62]

In 1687, Newton proved Kepler to be right, because the gravitational pull by the sun and the moon determines the tides.[63] Nevertheless, Galileo, too, was not far off the track, and within the context of what he knew of mechanics, his theory of the tides is a marvellous achievement. It shows an awareness of the relevance of acceleration, of inertia, and of resonance. Also in Newton's explanation, it is essential that the earth moves around the common centre of mass of the earth-moon system, and this motion being circular is accelerated. Hence, the aim of Galileo to prove the earth's motion by the tidal theory is also achieved if it is replaced by Newton's theory.

3.4 Cartesian physics

Between *c*.1620 and 1650, René Descartes was one of the leading "philosophers", regardless whether we take this word in the 17th-century meaning of "scientist", or in its modern meaning. His first published book, the *Discourse on Method* dates from 1637, but the appendices to this work, *Dioptrics*, *Meteors*, and *Geometry*, were written some time before. In particular the years he lived in the Dutch Republic (1629-1649) were very fruitful.

Descartes furthered the Copernican program of explaining motion by motion in several important ways. These concern the law of inertia, the mechanical properties of matter, the law of conservation of motion, and the problem of collision. Descartes was even more ambitious — he was convinced that with his method one could solve all problems of contemporary science. In the end, Cartesian physics was not very successful. But as a program to replace the Aristotelian scheme of explanation by a mechanistic one, Cartesian physics exerted a large influence. As such, Cartesian physics, though in many respects a *cul-de-sac*, constitutes an essential part of the Copernican revolution.

Inertia

The principle of inertia will only briefly be recalled. Galileo connected it with uniform circular motion, but his disciples took linear inertial motion for granted.[64] The Dutch scholar Isaac Beeckman (*c*.1620) distinguished two kinds of inertial motion — uniform rectilinear motion, and uniform rotation of a heavy body around its axis.

Descartes accepted the infinity of the universe, and therefore did not share Galileo's caution against rectilinear inertial motion. He posited the law that a body on which no force is acting, moves rectilinearly at constant speed.[65] This principle is now called "Newton's first law". It implies that circular motion has to be explained, only rectilinear uniform motion being accepted as a purely kinematic principle of explanation, an axiom of motion.

Matter and space

Just like Copernicus and Galileo, Descartes was strongly influenced by Plato, as can be seen from the pre-eminence of space and geometry in his work. Geometry, to which Descartes contributed in his deservedly famous *La Géometrie* (1637), is the paradigm of each science. The certainty provided by geometry is warranted by its method, and in order to arrive at the same level of reliability each science should proceed by the same method.

Descartes identified space with matter. All matter is space, and all space is material.[66] A vacuum cannot exist for it is unintelligible, and matter is infinitely divisible. For this reason, Descartes is usually considered not to have been an atomist.[67] This is right if atomism is characterized by the indivisibility of particles, moving in a void. However, for Descartes the divisibility of matter was a mathematical principle. On physical grounds he assumed the existence of particles, with a minimum dimension. He even distinguished three kinds of particles into which space-filling matter was differentiated.[68] Normal bodies are composed of coarse matter. Interplanetary space is filled with fine matter, and the pores in both are filled by the finest material, composing the sun and the stars, and responsible for the transmission of light. (On Descartes' view on light, see Sec. 11.3.)

As we have seen, Aristotle considered any substance a combination of form and matter. Essential and accidental properties determine any material body. Such properties were the essential properties of dry and moist, cold or warm, heavy or light, and accidental properties like colour and taste. Like most Copernicans, Descartes did not speak of essential and accidental, but of primary and secondary properties.[69] This distinction follows from the proposition that the real world is not necessarily the world as we perceive it. Primary qualities belong to objects as they really are. Secondary qualities such as heat or colour have no independent existence. The primary properties of matter are related to extension or motion — volume, quantity of motion, hardness, impenetrability, etc.[70] Other qualities are secondary, and should be reduced to primary, "mechanical" properties. An example is Descartes' reduction of magnetism to the motion of cork-screw particles (see Sec. 3.5).

If the Aristotelians talked about primary or manifest properties, they referred to sensory experience. Under Plato's influence, Galileo,[71] Descartes, and other Copernicans considered sensory experience to be secondary, in need of explanation on mechanical principles.

Impact
The introduction of the principle of inertia generates a problem unknown in Aristotelian physics — the problem of how motion can change. For Aristotle, celestial motion never changes, it being eternally circular, and sublunary motion simply ceases as soon as the body has arrived at its natural place. Violent motion ceases when the driving force no longer acts. Motion does not change, but *is* change.

After Galileo's introduction and Descartes' correction of the principle of inertia the question arose how motion can be started, halted, or changed in direction. Clearly this can only be done by some external force, for if no external force is acting on a body, it continues its motion. Galileo never posited this problem in his published work,[72] but Descartes did. It is the main problem of his physics.

Descartes wished to execute the Copernican program to explain motion by motion. Any movement can be changed only by another moving body. The only conceivable possibility for this is a collision between the two bodies. One can only speak of a force as far as it is the result of motion.

The problem of collision forms the heart of Cartesian physics. Descartes introduced "quantity of motion" as a measure of motion, operationally defined as volume (quantity of matter) times speed. Hence, his definition differs from the later definition of linear momentum in two ways. First, Descartes took volume as quantity of matter, because he identified matter with extension. Newton amended this by taking mass, operationally defined as volume times density, as quantity of matter (see Sec. 1.4). Next, Descartes considered velocity to be a scalar magnitude, *i.e.*, to be speed. This was corrected by Huygens, who assumed velocity to be a vector, as we now call it, having direction as well as magnitude.

Descartes assumed quantity of motion to be indestructible, because it is natural.[73] At the creation, God supplied the cosmos with a quantity of motion, which never changes. This is now called the *principle of conservation of momentum*. During a collision one body may transfer some quantity of motion to the other one, and this is the only conceivable way to change the motion of a body.[74]

Descartes elaborated these ideas into seven laws of impact.[75] With only one exception, these laws are contradicted by the results of experiments with colliding balls. Admitting this, Descartes observes that his laws concern circumstances which cannot be realized in concrete reality. They concern collisions between two bodies in a vacuum, and a vacuum is impossible. He says: "The proofs of all this are so certain, that even if our experience would show us the contrary, we are obliged to give credence to our mind rather than to our senses."[76] Again this demonstrates the influence of Platonic idealism in Descartes' views.

Rest and motion
From the laws of impact it can be derived that Descartes makes an

absolute distinction between rest and motion. Apparently he does not only use the concept of "quantity of motion", but also an independent concept, "quantity of rest", inertia. The quantity of rest dominates the quantity of motion if the body at rest is larger than the moving one.[77]

Galileo had denied the absolute difference between rest and motion. Rest is a state of motion, with zero velocity. In a beginning movement, the object starting from rest passes through all degrees of speed, until arriving at its final speed.[78] This is the foundation of the principle of relativity, whether Galilean or Einsteinian. If every movement has a relative character, there cannot be a fundamental distinction between rest and motion. When Huygens applied the principle of relativity to Descartes' laws of impact, all but one turned out to be false.

Descartes' distinction between rest and motion is necessitated by the problem of giving a foundation to the existence of bodies moving as a whole. The parts of the body move together with the whole, but are at rest with respect to each other. If Descartes would not have introduced the idea of rest, the idea of universal motion would have excluded the existence of extended bodies. By the distinction between rest and motion, Descartes tried to account for the mutual irreducibility of space and motion. Both spatial extension and motion are sustained by "forces" (and ultimately by God), whose effects are, respectively, inertia and quantity of motion.

Without admitting it in plain words, Descartes assumes some kind of absolute space, a space as seen by God. Elsewhere, Descartes contends that motion can only be relative.[79] This dilemma arises from his identification of space and matter. Because matter is the same as space, local motion as change of position is strictly speaking impossible. The only possibility to create motion in a plenum arises if spatial parts exchange their positions. Hence, real motion occurs in vortices, circular motion, returning into itself. Real vortical motion in a plenum is relative motion, and the non-existing idealized rectilinear motion in a void is absolute.

Because Descartes' problem of absolute and relative motion arises from his identification of matter and space, it differs profoundly from Newton's problem of absolute and relative space, time, and motion (see Sec. 3.6).

Planetary motion and gravity
According to Descartes the motion of all bodies, including the

celestial ones, takes place in a plenum. The sun turns around its axis, as was discovered by Galileo (among others) in 1612, and it drags along the surrounding matter, and hence the planets. The suggestion that the rotation of the sun causes the revolution of the planets was made by Kepler before Galileo's discovery.[80] Kepler estimated the period of revolution to be about three days, and he was disappointed to learn that the actual period is 30 days.[81] According to Descartes, if a planet turns around its own axis, it creates its own vortex, dragging around satellites like the moon.

Gravity is also explained from mechanical motion.[82] The vortical motion of matter around the rotating earth causes a centripetal motion of all bodies having density less than the whirling matter. Later on, Huygens thought he was able to demonstrate this effect in an experiment (see Sec. 4.2). Hence Descartes was the first to relate celestial motion with the motion of a falling body. Newton made the same connection, but in quite a different way (see Sec. 6.4). Before Descartes, Kepler compared gravity with magnetism, and magnetism with planetary motion, but never gravity with planetary motion. Kepler rejected the Aristotelian view of gravity,[83] and sustained Copernicus' view, saying that gravity is a mutual corporeal affection between cognate bodies tending to unite them.[84] Also Galileo never connected gravity with the "natural" motion of the planets.

3.5. Kepler's defection

Before Galileo, Kepler was the most prominent defender of Copernicus' views. He republished Rheticus' *Narratio Prima* (1540), the popular summary of Copernicus' system, together with his own *Mysterium Cosmographicum* (1597, 1621), and also in his other works on astronomy and optics, he paid tribute to Copernicus' views.

From 1600 to 1612 Kepler worked at Prague, then the capital of the German empire, first as Tycho Brahe's assistent, then as his successor as Imperial Astronomer, after Tycho's death in 1601. Kepler was charged with the description of the motion of the planet Mars with the help of Tycho's system. During the process, he found the first two laws of planetary motion, demonstrating that the planets move at a varying speed in elliptical orbits.

The sun in the centre
As a consequence, Kepler broke away from one of the fundamental ideas of Copernicanism — uniform, circular motion. Nevertheless,

Kepler always considered himself a Copernican, because he adhered to the idea of a moving earth.[85]

In another respect, he was even more Copernican than Copernicus. To call Copernicus' theory "heliocentric" is not entirely correct. The centre of Copernicus' system is the centre of the earth's orbit, slightly displaced from the sun's position. Also the other planets did not turn around the true sun, but about this centre, the "mean sun". This is especially relevant for the motion of Mars, because its orbital plane makes an inclination of nearly two degrees with the earth's orbital plane. According to Copernicus, the secant of these two planes passes through the mean sun.

Kepler corrected this. Right from the start of his study of the motion of Mars, he placed the sun in the centre, and hence the secant of the two orbital planes went through the sun. This led to a simplification of Copernicus' calculations, because now the inclination of Mars' orbit turned out to be a constant.[86]

In order to refer the observations to the sun, Kepler had to make a careful investigation of the motion of the earth, from which all observations are made. Nobody before him (not even Copernicus) had realized that. Kepler used an extremely ingenious and original procedure, consisting of "transporting" himself to the planet Mars, and observing the motion of the earth from that viewpoint. In order to eliminate the motion of Mars itself, Kepler took observations made 687 days apart, i.e., the period of Mars in its orbit.[87]

Force as a principle of explanation
Kepler's first two laws indicate exactly the failure of Platonic and Copernican models. The planetary orbits are not circular but elliptic, and the planetary motion is not uniform, but varies continually according to the area-law. Hence it should not be amazing that many Copernicans after Kepler rejected his results, and held to the uniform circular motion — for example, Galileo, Descartes,[88] and Huygens. It took quite a long time before other astronomers confirmed Kepler's discoveries,[89] shortly before 1687, when Newton published a theoretical justification.

Kepler and Newton shared many views.[90] Both realized that planetary motion was in need of a non-kinematical explanation, different from uniform circular motion. Even Descartes was aware of the need to distinguish motion from the "action, or inclination to move", but he reduced action to motion.[91] As we know, Newton found an explanation in the theory of gravity. Kepler's attempts were abortive. This is partly due to his lack of knowledge of the theory of mechanics, which still had to be developed by Galileo, Descartes, Huygens, and Newton, and partly due to his lack of insight into the concept of force, as we find it in Newton.

The Newtonian concept of force, which received its present precise meaning only after c.1850, can be traced back to Kepler. Because planetary motion turned out not to be uniform, it is not natural, it needs a force. Kepler supposed the planetary force to be proportional to its velocity. Like Aristotle, he assumed that without force there cannot be violent motion.[92] He conjectured that the sun exerts an influence on the planets, pushing them a-round in their orbital motion. Because a planet's velocity is largest if it is closest to the sun, Kepler assumed this "force" to be inversely proportional to the distance from the sun. This result he related to Archimedes' law of the lever. Hence, Kepler supposed his force to be a tangential one, directed along the planetary orbit. It was by no means attractive, *i.e.*, directed towards the sun.

Magnetism and the distinction between celestial and terrestrial physics
The idea of an attractive force was first contemplated with re-spect to magnetism. Like Galileo, Kepler greatly admired William Gilbert's book *De Magnete* (On the magnet, 1600).[93] Gilbert was a halfway Copernican, accepting the diurnal rotation of the earth, but not committing himself to its annual motion. Contrary to Peregrinus, who in 1269 ascribed the properties of the compass needle to the rotation of the *heavens*, Gilbert ascribed these to the rotating *earth*, which he considered a huge magnet.

Kepler subscribed to this view, and he assumed that the force exerted by the sun on the planets is magnetic.[94] He even assumed that the influence of the moon on the tides is magnetic. Galileo and Descartes rejected both ideas, because they wanted to explain motion by motion. Galileo executed this program with respect to the tides, whereas Descartes gave a mechanical explanation of magnetism. But Newton returned to the idea of Gilbert and Kep-ler, that magnetism is a *force* not to be explained by motion, but rather applicable as a new principle of explanation.

All Copernicans rejected the Aristotelian distinction between celestial and terrestrial physics. This could be done in two ways. The first is assuming that the same *laws* apply to both realms. This line was followed by Galileo, Descartes, and finally by Newton. The second way is to assume that the same *force* acts universally at the earth and in the heavens. This is Kepler's line, who considered magnetism in this capacity, and again Newton's, who demonstrated gravity to be the force determining planetary motion as well as the motion of falling bodies. "The force which retains the moon in its orbit is that very force which we commonly call gravity."[95]

The second line was first pursued by astrologers, starting from the assumption of a parallel development of celestial and terrestrial events. Kepler was the last Copernican to be sympathetic to astrology.

Magnetism and gravity as "occult" principles

The Aristotelians called qualities "occult" if these could not be reduced to the manifest, *i.e.*, sensorily observable qualities, such as hot and cold, moist and dry, hard and soft.[96] Also gravity and levity were manifest properties. Their key example of an occult property was magnetism.

The Cartesians considered properties "occult" if these could not be reduced to the clear and evident principles of mechanics. Hence they were proud of Descartes' achievement, the reduction of magnetism to the motion of cork-screw particles,[97] and the explanation of gravity by vortical motion. They objected to Newton's theory of gravity, with its inherent principles of attraction and action at a distance, which they considered "occult".

Needless to say that Newton rejected this view. For him, gravity was a "manifest" property, universally shared by all bodies, no less than their extension, hardness, impenetrability, mobility, and inertia.[98] He considered a property manifest if it can be measured. This partial return to Aristotelian empiricism shows that Newton took distance from Cartesian mechanism. He no longer reduced magnetism to mechanical motion, but measured its force.[99]

Force and pressure as static principles

The novelty introduced by Kepler and Newton concerns force as a *dynamic* principle, as a cause of change of motion. Aristotle too connected force with violent motion, but this motion was not variable.

Influenced by Archimedes, who studied the problem of the lever (and perhaps by the medieval scholar Jordanus Nemorarius), Tartaglia, Benedetti, Stevin, Galileo, Torricelli and Huygens developed the *static* principle of force.[100] The most important example of force was weight, the only kind of force considered by Archimedes, and problems concerning the centre of gravity were very prominent during the 17th century.

The study of Torricelli's void above a mercury column, and the rejection of the fundamental distinction between heavy and light objects, led to the insight that air has weight. Aristotelians explained the empty space in Torricelli's tube by the idea of a *horror vacui* — nature abhors the void. Descartes also rejected the possibility of a vacuum.[101] The Cartesians maintained that Torricelli's void was only empty of the coarse types of matter, but still contained the finest material responsible for the transmission of light. But Pascal argued that all experiments suggest that the space above the mercury column is empty. For him, experimental proof carried more weight than philosophical arguments.

The hydrostatics of Archimedes, Stevin, Galileo, Torricelli, and Mersenne, was perfected by Pascal in his *Traitez de l'équilibre des liqueurs, et de la pesanteur de la masse de l'air* (Treatise on the equilibrium of liquids and the weight of air, 1651-54, published posthumously in 1663.)[102] His most important concept was "pressure", now operationally defined as force per square metre. Pascal's theory is based on the axiom, now called Pascal's law, saying that in a static fluid, at the same level the pressure is everywhere the same. From this axiom one derives Archimedes' law, the properties of Torricelli's tube, and the phenomenon, predicted by Pascal, that the barometric pressure depends on height in the atmosphere.

The static principle of force, developed during the 17th century from Archimedes' Platonic ideas, must be carefully distinguished from the dynamic principle of force, which comes from Newton. Kepler's attempt to introduce a force in the explanation of planetary motion was still linked up with the Archimedean static principle of force, connected with the problem of the lever. Kepler recognized the problem, but did not solve it. Newton did.

3.6. Newton's dynamics

Isaac Newton's ideas on mechanics, gravity and planetary motion were shaped between c.1665 and 1685, during his years at Cambridge, quite long after the lifetimes of Kepler, Galileo, and Descartes, whose works he used and criticized.

Newton took pains to demonstrate that he was a true disciple of Copernicus (see Sec. 6.2), though not in the Platonic line of Galileo, Descartes and Huygens, but in Kepler's line. From Kepler he inherited the laws of planetary motion (which he fitted into his theory of gravitation), Kepler's respect for observations, and the idea of force as a dynamic principle. Later on, we shall discuss the theory of gravity (Sec. 6.4). At present, we restrict ourselves to the concept of force in the context of mechanics.

Newton's dynamics is presented as an introduction to the *Principia* (1687), which is mostly concerned with gravity and planetary motion. In merely 28 pages (in the English edition) Newton presents a summary of mechanics he is about to use. It contains operational definitions (see Sec. 1.4), several theorems, and a philosophical exposition of his ideas on space and time. He comments: "Hitherto I have laid down such principles as have been received by mathematicians, and are confirmed by abundance of experiments."[103]

Like Kepler, Newton considered force a principle of explanation independent of motion. This is the main difference between Kepler and Newton on the one hand, and the "pure mechanists", Descartes, Huygens, and Leibniz, on the other. The mechanists could only conceive of a force as an *effect* of motion.

Inertial force

The concept of force is by no means completely clear in Newton's work.[104] He distinguished several kinds of force, which we would no longer recognize as such. One is the *vis inertiae*, the force of inertia, also called *vis insita*, innate force. It is for any body proportional to or equal to its quantity of matter, now called its mass.

In Newton's conception this force resists change of motion. If besides the *vis insita* no other force is acting on the body, the latter moves uniformly and rectilinearly. We call this Newton's first law, but he took it from Descartes.[105]

External force

If an external force is acting on a body, its effect is inversely proportional to the body's mass, to the *vis inertiae*. Newton considered the case of a force acting during a short time. The effect of the "impulse" (force times duration) is the change of the quantity of motion (linear momentum), operationally defined by Newton as mass times velocity. This is Newton's force law, the second law of motion, nowadays better known by the operational formula, force equals mass times acceleration. This formula refers to a continually acting force, and is not explicitly given by Newton, but he uses it all the same.

Like Huygens, Newton considers velocity to have direction as well as magnitude. The same applies to force, for which he discusses the law of composition.[106] Hence, a force is present if speed, direction or both change.

Newton suggests that the force law is accepted by all his contemporaries, even by Galileo.[107] But Newton's interpretation of force as the cause of changing motion is probably original. We may consider Kepler as his predecessor, but Kepler had no quantitative expression available for the effects of a force.

The independence of the first and second laws

Sometimes people suggest that the law of inertia is superfluous, because it can be derived from the force law. If the net force on a body is zero, its acceleration is zero, hence its velocity is constant, they argue. The following objections may be put forward against this view.

First, both the common sense view and Aristotelian physics state that natural motion (*i.e.*, motion without an external cause) is accelerated. The first law denies this. Secondly, both common sense and Aristotelian science state with respect to violent motion that it ceases if the force ceases to act. Hence, without an unbalanced external force, a body remains at rest (apart from natural motion). The second law alone does not refute this view. Only in combination with the first law is the common sense view contradicted. These are historical arguments.

Thirdly, the force law states that a body moving under the influence of an external unbalanced force accelerates, but it does not specify with respect to what the acceleration is determined. The answer is, the acceleration is measured with respect to an inertial system. Apparently, this only shifts the problem, because now the question arises, with respect to what we can speak of an inertial system. The answer to this question is given by the first law: an inertial system is any body moving under the influence of no external force. (Newton defined inertial motion with respect to absolute space, but he admitted that this absolute motion cannot be measured; see below.)

This means that (as a matter of principle) the first law can be expressed like "There exist inertial systems". This axiom, expressing the mutual irreducibility of motion and interaction, is also valid in Einstein's theory of relativity.

Hence, the opinion that the law of inertia can be derived from the force law is mistaken.

Centrifugal or centripetal force

The distinction between force as *caused by* motion, and force as *cause of* motion is manifest in particular with respect to circular motion. Only since the introduction of linear inertia by Descartes, excluding circular inertial motion, has it become clear that circular motion is accelerated.

The first to investigate this problem was Huygens. He derived the correct formula for the acceleration, and as a true Cartesian, he introduced the *centrifugal* force as a *result* of circular motion. Newton, on the other hand, supposing that an acceleration needs a force, introduced the *centripetal* force as a *cause* of circular motion.[108] It is a real, physical force like a magnetic or gravitational force.

Since then, in Newtonian mechanics, centrifugal force is considered an *apparent* force, an "inertial" force. It is not a real force, because it only occurs in a non-inertial reference system, a rotating reference system, for example. Newton's laws presuppose inertial reference systems. Newtonians only admit of "inertial" forces, if introduced in a non-inertial reference system in order to be able to apply Newtonian dynamics. It is no more than a mathematical trick.

In particular, a "force" like the centrifugal force does not safisfy Newton's third law.

Action and reaction

Newton also ascribed the third law to others,[109] but that is no more than false modesty. The law was brand-new. It is the famous law of action and reaction: "To every action there is always opposed an equal reaction: or, the mutual actions of two bodies upon each other are always equal, and directed to contrary parts."[110]

Like Descartes, Newton assumed that the motion of a body can only be changed by another body. However, this change is not caused by the *motion* of the second body, but by the *force*, acting between them. Because this force has a reciprocal nature, the motion of the second body is changed as well. The law of conservation of linear momentum was for Descartes an axiom, but now became a theorem (a corollary) to be derived from Newton's laws.[111] Hence, force takes precedence over motion.

The third law may be considered the constitutional law of Newtonianism. It distinguishes Newtonian dynamics from Cartesian or Leibnizian mechanics.

The force-matter dualism

The three "axioms, or laws of motion" lie at the foundation of the Newtonian dualism of force and matter. Throughout his life,

Newton maintained an ambivalent position with respect to this dualism, because he was still under the influence of the Aristotelian idea of matter as being entirely passive.[112] It was difficult to accept that matter could be active. This activity comes through in the concept of inertia, the inertial force by which each body resists change of motion. Next, matter becomes active as a source of force, first of all the force of gravity.

Between 1700 and 1850, the flourishing period of Newtonianism, similar matter-force dualisms were developed in electricity (electric charge and Coulomb force), magnetism (magnetic force and pole-strength), and thermal physics (temperature difference as a force, and heat considered as matter).[113]

Force as a relation
In Newtonian mechanics, a force is considered a relation between two bodies, irreducible to other relations like quantity of matter, spatial distance, or relative motion. Though an actual force may partly depend on mass or spatial distance, as is the case with gravitational force, or on relative motion, as is the case with friction, a force is conceptually different from numerical, spatial or kinematic relations.

The conception of force as a relation must be contrasted with Descartes' quantity of motion or linear momentum, and Leibniz' *vis viva* (living force, operationally defined as mass times the square of speed), which were supposed to be properties of the moving body, not a relation between bodies.

In the 18th century, disciples of Descartes and Leibniz quarreled about the priority of momentum and *vis viva*. The Newtonian scholar d'Alembert demonstrated these concepts to be equally useful, momentum being the time-integral of the Newtonian force acting on the body, *vis viva* its space-integral.[114] But this compromise proposal was evidently unacceptable for both parties, because acceptance would imply the recognition of the priority of the Newtonian force.

Conservation laws
By the three laws of motion, force is attributed a higher priority than momentum. In the second law, force and momentum occur at the same level, and the first law mentions neither. But in the third law only forces are mentioned. Moreover, Newton showed that the law of conservation of linear momentum and Kepler's second law (the area law, an early expression of the law of conservation of angular momentum) can be derived from the third

law. In 1847, Helmholtz showed this to apply to the law of conservation of energy as well.

It was only later realized that this derivation was subject to a rather severe condition. It is the assumption that all forces can be reduced to so-called "central forces", working between point-masses. This condition was acceptable for most Copernicans and the later Newtonians, because it fit very well into their atomistic views. But it was far less acceptable for the Cartesians, who rejected atomism in favour of the identification of matter and space.

Until the end of the 19th century, the third law maintained its priority over the conservation laws. Only in the field theories of Maxwell and Einstein, implying a return to Cartesian views of matter and space, it turned out that the above mentioned condition is too narrow. At present, physicists prefer to reverse the situation. Newton's third law is shown to be a consequence of the law of conservation of momentum, under certain conditions. Hence, nowadays the conservation laws have a higher priority than Newton's laws of motion, and the concepts of energy and momentum are more important than the concept of force.[115]

This does not mean that conservation laws were absent in Newtonian mechanics and physics. Related to the material side of the matter-force dualism, several conservation laws were developed after Newton's death — the laws of conservation of matter in chemistry, of electric charge in electricity, of magnetic pole-strength in magnetism, and of heat (caloric) in thermal physics.[116]

Absolute and relative space, time, and motion

One quarter of Newton's summary of mechanics is devoted to a "scholium" on space, time and motion.[117] He did not intend to give definitions of these concepts, "as being well known to all." He wanted to make a distinction between "absolute" and "relative" space *etc.* It should be emphasized that his use of the term "relative" differs from ours. By "relative time", for instance, Newton meant time as actually measured by some clock. Some clocks may be more accurate than others, but in principle no measuring instrument is absolutely accurate. Hence, Newton was looking for a *universal standard* of time, space, and motion, independent of measuring instruments. In this respect, it should be admitted that he was more successful in posing the problem than in solving it.[118]

Before the Copernican revolution this problem did not exist.

Aristotle defined time as the measure of change, but because his physics was never developed into a quantitative theory of change, this conceptual definition never became operational. Galileo was the discoverer of the isochrony of the pendulum. Its period of oscillation depends only on the length of the pendulum, and is independent of the amplitude (if not too large) and of the mass of the bob. Experimentally, this can be checked by comparing several pendulums, oscillating simultaneously, but this does not prove that successive periods of the same pendulum are equal.

In 1659 Huygens derived the pendulum law making use of the principle of inertia, but apparently he did not see the inherent problem of time. Like Aristotle and Galileo, he just assumed the daily motion of the fixed stars (*i.e.,* the diurnal motion of the earth) to be uniform, and thus a natural measure of time. But Newton's theory of gravitation applied to the solar system shows that the diurnal motion of the earth may very well be irregular; it is a "relative" measure of time in Newton's sense.

The problem of "absolute" time, space and motion is most pregnantly expressed in Newton's first law, the law of inertia: "Every body continues in its state of rest, or of uniform motion in a right line, unless it is compelled to change that state by forces impressed upon it." "Uniform motion" means that equal distances are traversed in equal times. In the context of our idea of "irreducible modes of experience", we take the law of inertia to express the mutual irreducibility of the spatial and kinematic modes. This means that the "absolute" standard of time is operationally defined by the law of inertia itself. The accuracy of any actual clock should be judged by the way it confirms this law. The law of inertia is a genuine axiom, because there is no experimental way to test it. It is more than a definition, as far as the irreducibility of space and motion is more than a matter of convention.

However, Newton did not follow this path. The only way he saw to solve the problem was to postulate an absolute clock: "Absolute, true and mathematical time, of itself, and from its own nature, flows equably without relation to anything external", and similarly: "Absolute space, in its own nature, without relation to anything external, remains always similar and immovable."

Newton admits that the velocity of some inertially moving body can never be determined with respect to this absolute space, but he maintains that non-uniform motion with respect to absolute space can be determined experimentally. He illustrates this view with an experiment on a rotating pail of water.[119] Hang a pail of water on a rope, and make it turn. Initially, the water remains at rest, and its surface horizontal. Next, the water begins rotating, and its surface becomes concave. If ultimately, the rotation of the pail is arrested abruptly, the water continues its rotation, maintaining a concave surface. Newton concludes that the shape of the surface is determined by the absolute rotation of the fluid, independent of the state of motion of its immediate surroundings. Hence, observation of the shape of the surface allows us to determine whether the fluid rotates or not. In a similar way, Foucault's pendulum experiment (1851) allowed him to prove the earth's rotation without reference to some extraterrestrial reference system, such as the fixed stars.

Newton's views on space and time were discussed by Leibniz and Samuel Clarke (acting on behalf of Newton), who in 1715-16 each wrote five letters.[120] Against Newton, Leibniz contended that space and time merely constitute relations between events. Denouncing absolute space and time, he said that only relative space and time are relevant; but it is clear that "relative" now means something different than it did for Newton.[121]

Leibniz' opinions were repeated by Ernst Mach in the 19th century, who in turn influenced Einstein. Mach even denied the conclusion drawn from Newton's pail experiment.[122] He said that the same effect should be expected if it were possible to rotate the starry universe instead of the pail with water. The rotating mass of the stars would have the effect of making the surface of the fluid concave. This means the assumption that the inertia of any body is caused by the total mass of the universe.[123] However, it has not yet been possible to find a mathematical theory giving the effect predicted by Mach. Mach's principle, stating that rotational motion is just as relative as linear uniform motion, is therefore as yet unsubstantiated. Both in Newtonian and Einsteinian mechanics as now conceived, it is taken that rotational motion has an absolute character, not relative to absolute space, but relative to a local inertial system. Rotational motion of an extended body is still in need of physical explanation, because it is only possible if the parts of the body are kept together by some internal or external force. Experiments like Newton's or Foucault's can only demonstrate some system to be non-inertial. They cannot prove any system to be inertial, however.

Motion and interaction as mutually independent principles of explanation

In Newton's work force is the most important concept besides matter. This may be called the strongest rupture with the mechanists, who wanted to explain motion by motion. For Galileo and Descartes, matter was characterized by quantity, extension, and motion.[124] Motion could only be caused by motion.[125] Newton emphasized that perceptibility and tangibility were characteristic of matter just as well. The property of matter to be able to act upon things can not be grounded on extension alone.[126] Newton introduced a new principle of explanation, now called interaction. Besides numerical, spatial and kinematic relations, interactions turn out to be indispensable for the explanation of natural phenomena.

This is of great importance. Galileo and Descartes showed motion to be a principle of explanation independent of the numerical and spatial principles. This led them to the law of inertia, Newton's first law. Descartes *hypostatized* this new principle, because he assumed that all natural phenomena should be explained by motion as well as matter, conceived to be identical with space. Newton *relativized* this principle, by demonstrating the need of another irreducible principle of explanation, the physical principle of interaction.[127]

The numerical, spatial, kinematic, and physical relations serve as principles of explanation, because they are modes of human experience as well as modes of being. Newton's introduction of force as a dynamic principle opened up the physical aspect, but only just. For, as a Copernican inspired by the idea that the earth moves, his real interest was in the explanation of all kinds of motion, including accelerated motion. The full exploration of the physical aspect did not occur in the Copernican era, but in its successor, the period of Newtonianism (*c.*1700-1850).

Although this Copernican commitment partly justifies Dijksterhuis' view that Newton fulfilled the "mechanization of the world picture", the distinction between Cartesianism and Newtonianism is important enough to shed some doubt on this view. It is not improbable that Dijksterhuis considered the Copernican era too much from the viewpoint of late 19th-century mechanism, which included a short revival of Cartesianism.

The mutual irreducibility of the four modes of experience mentioned earlier is most pregnantly expressed in the principle of relativity and its consequence, the objectivity of physical interactions with respect to time, space, and motion (see Sec. 5.4).

4. The solution of problems

4.1. The various functions of the statements in a theory

In the preceding chapter we discussed the relevance of the Copernican revolution with respect to the first four "irreducible modes of experience", as we called them — the numerical, the spatial, the kinematic and the physical ones. The next, obviously, is life, the biological mode of being. It has never been connected with Copernicanism, but we shall be concerned with this aspect of human experience, in our study of theories, their structure and functioning. Life includes growth, metabolism, generation. For theories, this means the growth of knowledge, the transformation of data, the generation and solution of problems. This is no more (and no less) than an analogy. Life also includes the internal differentiation of the parts of a living body, according to their internal functioning. Likewise, in a theory, defined as a deductively ordered set of true statements, we distinguish propositions having various functions.

In Chapters 1 and 2 we discussed four stages in the use of theories, to wit, identification (a planet is a wandering star, the morning star is identical with the evening star), description (of the planetary motion through the zodiac, of backward motion), prediction (of solar or lunar eclipses), and explanation (of retrograde motion, for example). In the present chapter, we discuss the fifth stage, the solution of problems.[1]

Functional differentiation
Neither our provisional definition of a theory (Sec. 1.3), nor its ensuing criterion whether a statement does or does not belong to a theory, differentiates between the functions of the statements which form the elements of the theory. But we already found occasion to distinguish between universal or law statements, and existential or factual statements (Sec. 1.5).

In the present section, we shall distinguish statements according to their function in the deductive process carried out in the theory. Whether a statement or proposition is a definition, an axiom, a presupposition, a lemma, a datum, a theorem, a problem, or a hypothesis is mostly determined by the theoretical context, and may be different in another context.[2] As in a living organism, the whole determines the functioning of the parts.

Definitions

It is by a definition that a concept or property is introduced into a theory. Because concepts and properties are not statements, they cannot be elements of a theory, that is, a set of statements. Usually, a definition does not fully determine the extension and intension of a concept, because these are to be developed in the theory itself. In this respect, definitions are theory-dependent, or theoretical.

Definitions concern general or class concepts (*e.g.*, "planets"), as well as particular ones (*e.g.*, "the moon"). A peculiar kind of definition is the operational one (see Sec. 1.4).

Axioms

Axioms are law statements, supposed to be true within the context, but not provable by the theory. In Newton's theory of gravity, the inverse-square law is the fundamental axiom. In Newton's theory of planetary motion, the central axiom is the assumption that the gravitational force is the only one acting between the bodies of the system.

No theory can function without one or more axioms, without law statements.[3] Every theory is *characterized* by its axioms.[4] Hence, Newton's theory of planetary motion differs from Descartes', because each starts from different axioms, but Kant's theory is only an extension of Newton's.

However, the same theory can be axiomatized in various ways, either because axioms and theorems can change parts within a theory, or because the same physical law allows of various mathematical formulations.[5]

More about axioms appears in Sec. 4.4.

Presuppositions

Presuppositions are law statements supposed to be true, which are borrowed from other theories. In Newton's theory of gravity, suppositions are the laws of mechanics (*i.e.*, the three laws of motion and a number of theorems derived from them), and several branches of mathematics.[6]

Presuppositions are indispensable for a theory, but not characteristic. They are exchangeable. We may exchange one set of presuppositions for another one without changing the theory significantly. Newton's theory of gravity does not essentially change if
we exchange Newton's geometrical methods of proof by those of
the calculus, or if we replace the third law of motion by the laws
of conservation of energy and linear momentum, or if we replace
Newton's version of Kepler's second law (the area law) by the
law of conservation of angular momentum. A theory cannot function without these presuppositions, but is not severely tied to
them.[7] More will be said about presuppositions in Sec. 4.5.

Data

Data are factual statements, often derived in other theories, and
are taken to be true within the context of the theory. The truth of
a datum strongly depends on the context. In one theory a statement
may be a problem, or the solution of a problem, in another theory
it may be a datum, accepted truth. Outside the context of a theory,
a datum may be considered false, or at most approximately true.
Even in the same theory, we often use successively, but never
simultaneously, data contradicting each other. For instance, in
Newton's theory of planetary motion, planets are first considered
to be points, next to be perfect spheres, although both statements
are false and contradictory (see Sec. 1.3).

In theories of motion, data often consist of the positions and
velocities of the moving bodies at a given instant. For this reason,
data are often called "initial" or "boundary conditions", or "information."[8] A particular set of initial conditions may constitute a
"model" (see Sec. 6.2).

Even more so than presuppositions, data are exchangeable in
a theory.[9] We can even exchange a datum by its logical negation.
But data are indispensable for the solution of most if not all
problems.

More is said about data in Sec. 5.3.

Theorems

Propositions or theorems are factual or law statements which are
derived (deduced) within the theory. Their truth is thereby
established only as far as the axioms, presuppositions, and data
from which they are derived are accepted to be true. A theorem
will often be the solution of a problem.[10]

Hypotheses

Hypotheses have the same function as axioms or data, and nowadays philosophers do not always care to distinguish between, *e.g.*, axioms and hypotheses. In Chapter 11 we shall see that during the Copernican revolution, hypotheses had a lower status than they have now. The truth of a hypothesis was not taken for granted.[11]

A hypothesis may be a tentative solution of a problem, or a statement from which a solution can be derived. It is introduced to test its fertility, its problem solving capacity.

In an absolute sense, every axiom, theorem or datum is a hypothesis,[12] but such a radical view, though very common, is not very useful if we want to distinguish between various functions of the elements of a theory. Within the context of a theory it is possible and fruitful to distinguish between statements the truth of which is not doubted by the users of the theory, and statements whose relative truth value is in question. Hence, we take hypotheses to be tentative, temporal, and often speculative elements of a theory.

Problems

Problems are not statements. They have a logical structure different from concepts, statements or theories. Apparently, they have biological characteristics, besides logical ones. Problems are born, they flourish, and bear fruit, they generate new problems, and they die. Only the solution of a problem is a statement, a theorem. The most simple form of a problem is "to prove the following statement", which is the form of an instruction. However, in many cases a problem can only be formulated in this way, if its solution is known. Usually, unsolved problems will have a more "open" form.

As long as a problem is not solved, it cannot be decided whether its solution satisfies the criterion according to which a statement does or does not belong to the theory. Hence it is not always obvious whether a theory should be able to solve a problem, that is, whether the problem "belongs" to the theory. For instance, before 1577, comets were considered meteorological phenomena, occurring at the periphery of the sublunary spheres, and problems concerning comets belonged to meteorology. Only when Tycho Brahe demonstrated their distance to be larger than the moon's, comets became celestial phenomena, and their motion became a problem for astronomical theories. But still in 1618, Galileo expressed his doubt of Tycho's proof (see Sec. 7.3).

It should be observed that problems may be quite meaningless outside the context of a certain theory. For instance, the problem of determining the relative distances of the planets to the sun from the observed retrograde motions, which is solvable in Copernicus' theory (see Sec. 2.5), is meaningless in Ptolemy's theory, in which the relation of the retrograde motion to the sun's is merely accidental.

A problem may be called *"theoretical"* if it can only be posed and eventually solved within a certain theoretical context. Hence, the problem mentioned above is "theoretical" with respect to Copernicus' theory. But the related problem, to determine the *absolute* distance of the earth to the sun is not, for its value is determined "experimentally", even though in this "experiment" theories (other than Copernicus') must be applied. By the way, the first reliable determination of this distance was only made at the end of the 17th century. All earlier estimates were far too small.

Obviously, this is only one way to define "theoretical" problems. Another way is to look at the reason why we want to solve a problem. Now, a "theoretical" problem is solved in order to develop the theory, or to show its superiority over competing theories, whereas a "practical" problem is solved in order to meet some extratheoretical need.

Input-output scheme
The above classification of statements is probably not exhaustive, but suffices for our purpose, to proceed with our analysis of the structure and functions of a theory. Implicitly, we distinguished the *core* of a theory from its *periphery*. A theory has a relatively small *nucleus* of statements characteristic of the theory — its axioms and a small number of theorems. This nucleus is surrounded by a vast *cloud* of other statements — an unspecified number of presuppositions, a potentially infinite amount of data, and an unknown number of problems.

This scheme has an input-output character, like a living system. The input consists of a problem together with some data considered to be necessary to solve the problem. The output (the fruits of the theory) consists of the solution of the problem. In the above example, the input is the problem to determine the relative dimensions of the planetary orbits, and the data are the observed sizes of the retrogradations. The output, the solution of the problem, consists of data, which can be used as input values for another theory, for instance, Kepler's model of the solar system. Hence, if a theory functions well, it contributes to the growth of our knowledge.

Data, presuppositions and even problems may have their origin in other theories, and the solutions of problems may be useful in still other theories. Hence, each theory forms part of a network of theories (see Chapter 5). Stripped from data, presuppositions and problems, a theory is no more than a deductive *scheme* containing "blanks" or "variables."

4.2. Normal science

Recently, both Karl Popper and Thomas Kuhn have stressed that theories should solve problems, and thus contribute to the growth of our knowledge.[13] Popper points to the biological character of problem solving, and Kuhn has become famous because of his theory of "paradigms" and "normal science." Though we have little use for Kuhn's views on scientific revolutions (see Sec. 4.5), we consider his discovery of "normal science" to be of outstanding value.

Trial and error
According to Popper, the main if not the only method of solving problems is "trial and error", an idea drawn from Darwin's theory of evolution.[14] He says: "Human thought tends to try out every conceivable solution for any problem with which it is faced...by the method of trial and error. This, fundamentally, is also the method used by living organisms in the process of adaptation."[15] "If the outcome of a test shows that the theory is erroneous, then it is eliminated; the method of trial and error is essentially a method of elimination. Its success depends mainly on three conditions, namely, that sufficiently numerous (and ingenious) theories should be offered, that the theories offered should be sufficiently varied, and that sufficiently severe tests should be made. In this way we may, if we are lucky, secure the survival of the fittest theory by the elimination of those which are less fit."[16]

Popper's view is doubtless attractive because of its radical simplicity. But it is hardly probable that any professor would accept a student literally working on a scientific problem according to Popper's recipe.

Paradigms
In his influential book, *The Structure of Scientific Revolutions*

(1962), Thomas Kuhn identifies "normal science" with the problem-solving stage in the history of any field of science, which follows the acceptance of a "paradigm."[17] For examples, Kuhn points to Aristotle's *Physics*, Ptolemy's *Almagest*, Copernicus' *Revolutionibus*, and Newton's *Principia* and *Opticks*.[18] Unfortunately, Kuhn's concept of a paradigm was at first highly ambiguous,[19] but in his *Postscript — 1969*, he stated that the term "paradigm" is used in two different senses. "On the one hand, it stands for the entire constellation of beliefs, values, techniques, and so on shared by the members of a given community. On the other, it denotes one sort of element in that constellation, the concrete puzzle-solutions which, employed as models or examples, can replace explicit rules as a basis for the solution of the remaining puzzles of normal science."[20]

We have some doubt with respect to Copernicus' *Revolutionibus*. Nobody tried to solve problems according to Copernicus' methods, except Kepler, and he failed. Also in the first sense, the *Revolutionibus* hardly counts as a "paradigm." It was the belief that the earth moves which made someone a Copernican, but most Copernicans did not bother to read Copernicus' book. But the other four examples mentioned above certainly satisfy Kuhn's definition of a paradigm, at least in the second sense. For many centuries, Aristotle's *Physics* determined the character of an explanation. Ptolemy's *Almagest* showed how to calculate planetary motions, Newton's *Principia* how to conduct theoretical science, and his *Opticks* how to do experimental work in physics and chemistry.[21]

These examples, however, enable us to shed doubt on Kuhn's suggestion that a paradigm has an exclusive character such that anybody who does not accept it, excludes himself from the scientific community. During the middle ages, Aristotle's philosophy was continually challenged by Platonists, and Ptolemy's heterocentric system was challenged by the Aristotelian homocentric system (see Sec. 2.1). Copernicans warred simultaneously with adherents of Ptolemy and of Tycho Brahe. Newtonians had Cartesians as competitors.

Even a single scientist may work simultaneously according to several paradigms. For example, between 1600 and 1606, Kepler tried to solve the problem of the motion of Mars according to Ptolemy's, Copernicus', and Tycho's methods. He failed, but his own solution (the laws of planetary motion) never became a paradigm, as we have seen (Sec. 3.5).

Textbook science

Partly, normal science is the practicing of science in an educational context.[22] After the acceptance of a paradigm, textbooks are no longer different with respect to their contents. At most they differ by the choice of examples, their didactics, or their lay-out. Textbooks only present the accepted views of the scientific community. Alternatives are not mentioned at all, or at best as examples to be avoided. The structure of a textbook is authoritative. The theory, the only true doctrine, is presented as incontestable truth.

At the end of each chapter, problems are given in order to train the student to solve problems according to the method suggested by the book. Kuhn stresses that this is the only way for a student to understand the meaning, for instance, of Newton's force law. Only if a student has learned enough he may try to solve problems which have never been studied before. Many students never arrive at this stage. Independently solving a completely new problem is research leading to a Ph.D., the highest degree a university is able to grant. By solving problems, a student proves his ability to become a scientist. According to Kuhn, normal science is largely concerned with puzzle-solving.[23]

However, normal science is more than science in an educational context. It also means the development of a theory, without doubting its fundamental axioms, its nucleus, as defined above. This is done by solving problems.

Usually, problems cannot be solved by a given set of axioms, theorems, presuppositions, and data. New theorems have to be derived, presuppositions must be developed, data criticized, and new data must be found. Collecting data is only meaningful in the context of a theory, with the aim to solve a problem, and each problem needs its own set of data.

Hence, normal science is by no means as uncritical as is suggested by the textbook metaphor, even if the nucleus of the theory remains unquestioned. Therefore, the criticism leveled by Popper and others concerning the supposed uncritical character of normal science seems to be unwarranted.[24]

Because in normal science the solution of problems is central, it is largely responsible for the growth of our knowledge.

The theory-dependence of problems

Whereas a theory is characterized by its axioms, a field of scientific research is characterized by its problems. According to Kuhn a social group of scientists is determined by their sharing a paradigm.[25]

It can hardly be denied, however, that astronomy is a field of science, practiced by astronomers, who during important historical periods were guided by widely different paradigms — for instance, the Copernican versus the Aristotelian, or the Cartesian versus the Newtonian. Galileo's *Dialogue* (1632) shows conclusively that adherents to competing paradigms were very well able to communicate their problems and to discuss each other's solutions.

This means that problems are partly autonomous with respect to theories. Kuhn denies this. Because he assumes that a paradigm not only determines how problems should be solved, but even determines the problems themselves, he thinks that adherents of competing paradigms cannot even understand each other's problems.[26]

However, problems are not completely independent of a theory. The same problem may appear differently in various theories. In Ptolemy's theory it is a problem to describe retrograde motion with the help of fictitious deferents and epicycles. Copernicus explained retrograde motion as a logical consequence of his theory. The non-observability of the stellar parallax is a problem in Copernicus' theory, but not in Ptolemy's, who, however, would have had no trouble to understand it.

Anomalies

Problems having an easy solution in one theory may be insoluble in an alternative theory. For example, Aristotle's theory of crystalline spheres surrounding the earth explains why the moon always turns the same face to the earth. No Copernican theory could explain this. Descartes' theory of vortical motion (Sec. 3.4) easily explains why all planets turn around the sun in the same direction, and more or less in the same plane. Moreover, they rotate about their axis in the same direction, and this also applies to the sun and the moon.[27] This problem has never been solved in Newton's theory.

Thus, within a theory we encounter unsolved problems besides solved problems and anomalies. We call an unsolved problem an *anomaly* only if it can be solved by a competing theory.[28] Hence, anomalies are theory-dependent. Einstein's general theory of relativity (1916) is able to solve the problem of the perihelion motion of Mercury. Before, it was an unsolved problem in Newton's theory, but since 1916, it is an anomaly in that theory.

For a discussion of the merits of rivalling theories, anomalies will be relevant. A theory containing anomalies is suspect, and the removal of an anomaly will count as a triumph over competing theories.

Christiaan Huygens

It turns out to be difficult to allot the Dutch scientist Christiaan Huygens his right position in history, amidst the giants of the Copernican revolution, Copernicus, Kepler, Galileo, Descartes and Newton. Apparently, Huygens contributed little to the revolutionary character of 17th-century science. But he was considered the foremost mathematician of his generation, and an important physicist and astronomer.

Perhaps he may be characterized as a prototype of a normal scientist. Like Kepler, Galileo and Descartes before him, and Newton after him, he solved many problems, but he does not belong to the architects of Copernicanism and mechanism, to which views he adhered.

As a *mathematician*, Huygens followed Descartes in his analytical geometry, and Torricelli in calculations of centres of gravity. Through Fermat and Pascal he became interested in problems of chance. In all three fields he excelled, but he contributed nothing to the development of the calculus, which he left to his younger contemporaries, Newton and Leibniz.

As an *astronomer*, he improved the telescope by the invention of a new ocular and the micrometer. In 1655 he discovered the first moon of Saturn. He explained the unusual appearance of Saturn, first observed by Galileo in 1610, as being caused by a ring around this planet. His *Systema Saturnium* (1659) was the most important work on telescopic astronomy since Galileo's.[29]

As a *physicist*, Huygens discovered the laws of pendulum motion and of centrifugal acceleration in uniform circular motion, and he explained the double refraction of light in Icelandic crystal. In order to achieve this he developed a theory of light from an extant theory by Descartes. He also improved Descartes' laws of collision (see Sec. 3.4), realizing that the law of conservation of quantity of motion can only be valid if this quantity has direction as well as magnitude. Moreover, he applied the principle of relativity to the problem of impact.

As a *designer* Huygens became famous because of the pendulum clock (1657) and the spring balance clock (1674-75). He is called the inventor of the pendulum clock, not because he discovered that any new principle applied, but because he was the first to bring various principles together, and to build a working instrument. Initially it was intended as an instrument for the determination of longitude at sea, for which a pendulum clock is obviously not well fitted. This drawback stimulated Huygens to devise the spring balance, lying at the foundation of the ship's chronometer.

Huygens combined experimental with theoretical work. In order to investigate Descartes' theory of gravity, he put pieces of sealing wax into a pail of water (1668). If the pail were rotating, the pieces of wax moved to the periphery. But if the rotation were interrupted, all pieces moved to the centre, "...in one piece, which presented me the effect of gravity."[30] By this he argued that gravity is a centripetal force, caused by the vortical motion of matter around the rotating earth. One objection against Descartes' theory was that the density of the imperceptible whirling matter would have to be larger than the density of the bodies falling to the earth. Moreover it is difficult to understand why gravity is directed to the centre, rather than to the axis of the earth's rotation. In an ingenious way, Huygens sought to meet these objections. He published his theory, developed in 1667, only in 1690, three years after the much more successful theory of Newton, which Huygens admiringly but critically discussed.

Although, especially when growing old, he experienced trouble in accepting really new developments, Huygens was a versatile and fertile problem solver. His great merits were recognized by Colbert, the minister of Louis XIV, who wanted Paris to become the cultural and political capital of the world. He persuaded Huygens to become the leader of the *Académie Royale des Sciences*, founded in 1666. With some interruptions, Huygens worked there till 1681 on his own research, stimulating others.

Although he may have had few revolutionary ideas, Huygens was by no means a slavish imitator of others. He adopted ideas from Descartes, but criticized them too, developing them in his own way. Newton's work he admired, but he did not share his ideas. It may be surmised that Huygens was too critical to be original. Yet, this is typical for normal science, as we understand it. From Kuhn's description, normal science can too easily be concluded to be uncritical, bound up as it is to a paradigm. But if we consider Huygens as a prototype of a normal scientist, we must assume that slavish adherence to a paradigm is not necessarily prominent in normal science.

Mature normal science should be distinguished from textbook science, described above. Typical for normal science is the central position of problem solving, as well as a critical view of ideas and theories suggesting solutions. A normal scientist worth his name should feel free to make a choice between various paradigms available, even if he lacks the creativity to invent a new paradigm, a new method of solving problems.

4.3. The generation of problems

"Apparently all vertebrates have a capacity for taking notice of problems of some kind...yet man alone invents new problems."[31] The fertility of a theory can be judged by its capacity to *solve* problems, as well as by its capacity to *generate* new problems.[32] Copernicus' theory was accepted by many people not because it solved existing problems, but because it challenged creative people to solve new problems, connected with the idea of the moving earth.

In general, it is not the theory itself, but the solution of a problem with the help of the theory which generates a new problem. However, it should not be overlooked that, ultimately, it is neither a theory nor a solved problem, but some *scientist* who invents a new problem. It is suggested to him by the theory and its consequences.

The merit of great scientists like Kepler, Galileo, Descartes and Newton is not first of all the capability to solve problems, but to pose the right questions at the right time, the recognition of relevant problems.[33]

Retrograde motion and parallax
The question which was first, a problem or a theory, is like the question of the chicken and the egg. One of the oldest astronomical exercises, to explain retrograde motion, only became a problem after the acceptance of Plato's theory which assumed all celestial bodies to move uniformly in circular orbits.

Copernicus' solution of this problem generated the problem of stellar parallax. According to his theory, not only the planets but all stars should display an apparent motion caused by the annual motion of the earth. Only in 1838 did F. W. Bessel actually observe this so-called parallax. Up till then, Copernicus and his adherents had to explain why parallax was unobservable, by pointing to the limited accuracy of observation results, and by an *ad-hoc* assumption. This hypothetical datum assumes that the stars are very far away, compared to the distance from the earth to the sun.[34]

Therefore, with respect to the universe at large, Simon Stevin observed that there is little sense in distinguishing between a heliocentric and a geocentric system.[35] If the size of the terrestrial orbit is negligible, it can be held that the earth is still the centre of the universe. It is a more relevant distinction that the sun is the centre of the planetary system, and that the earth is moving around it.

The earth's motion

Copernicus' theory generated far more problems than it solved, especially because the earth's motion cannot be experienced. Would not the air and the clouds fall behind the earth? Would not a cannon ball fired to the west fly much farther than a ball fired to the east? Would not an object falling from the top of a tower show a deflection to the west, instead of falling vertically down? Would not people at the equator be hurled off the earth?

Many of these problems were faced by Copernicus (and even by Oresme, in the 14th century). Copernicus stressed that water and air rotate together with the earth.[36] He refuted Ptolemy's argument that a revolving earth would break apart, by stating that the earth's rotation is natural, a consequence of its spherical shape. Hence it cannot have any violent effects, for a natural motion destroying the nature of the moving body would be a contradiction in terms.[37]

Later on, Bruno, and especially Galileo, took pains to refute the arguments against the motion of the earth.[38]

The secondary light of the moon

The Copernican theory generated the problem of proving that the earth is not fundamentally different from the other planets. Galileo used his discovery of the phases of Venus to show that the earth, the moon, and Venus share the property of reflecting the light of the sun. For the moon, this property was generally acknowledged. The observed phases of Venus are similar to the phases of the moon, and can only be explained by assuming Venus to be a light reflecting body instead of a primary source of light like the sun or the fixed stars.

Next, Galileo argues that the earth also reflects the light of the sun. He does so by pointing to the so-called secondary light of the moon, occurring shortly before or after new moon, when alongside a small sickle the "dark" part of the moon is perceptible.[39] Galileo also explains why this phenomenon is not observable at first or last quarter.

Probably, Galileo did not know that the same explanation had already been given by Leonardo da Vinci and by Maestlin.[40]

Kepler's laws

Kepler's rejection of uniform circular planetary motion generated the problem of force — the problem why planets change their speed, and deviate from their circular orbits (Sec. 3.5). This problem was solved by Newton.

Initially, in his *Mysterium Cosmographicum* (1597) Kepler assumed an animistic view, supposing each planet's motion to be conducted by a soul. But in a footnote added in 1621, he says that everywhere the word "soul" should be replaced by "force", the moving soul of the planets.[41]

Impact

Descartes' identification of space and matter implied the existence of only one possible mode of interaction, action by contact. Combined with his view on the indestructibility of motion, this led him to an investigation of the problem of impact (Sec. 3.4).

Shortly after its foundation (1662), the Royal Society of London organized a contest about the problems of impact, and the prize was won by three contestants, Huygens, Wallis, and Wren. In fact, each of them solved a different problem. Huygens proceeded on the road of Descartes, by applying the law of conservation of motion and using a coordinate transformation. This technique, invented by Huygens, depends on the principle of relativity of motion, and later became a powerful expedient for the solution of many related problems.

Wallis and Wren applied the law of conservation of *vis viva*. Wallis studied inelastic impact, Wren and Huygens elastic collisions.

Newton

In 1685 Newton was challenged by Halley to solve the problem of whether an elliptical orbit would agree with an inverse-square law concerning the attraction of the planet by the sun.[42] Newton solved this problem, but the ensuing theory generated various new problems.

It generated the problem of the disturbance of the planetary motion by other planets. It generated the problem whether a planet moves around a resting sun, or whether both sun and planet move around their common centre of mass. The theory generated the problem whether the gravitational force exerted by a spherical body of given mass depends on the body's radius.[43] It generated the problem about the influence of the sun and the moon on the tides,[44] and of the sun on lunar motion. It generated the question whether the motion of a falling apple is of the same nature as the moon's motion. It generated problems concerning the shape of the earth and its influence on the acceleration of falling bodies, besides the latter being influenced by the earth's rotation. It led to the question whether gravity and magnetism are the same force.

Between 1685 and 1687 Newton solved most of these and many other problems, of which the first book of the *Principia* (1687) mentions 48, and the third 22. The most obstinate problem, concerning the lunar motion, occupied Newton till shortly before 1700, when he gave it up. It was only approximately solved several decades after his death.

The reinforcement of a theory

Its problem-generating capacity makes a theory an instrument for the investigation of nature. Generating and solving problems constitutes a great deal of the history of science. History of science is the history of its problems.[45]

Because a theory generates new problems it may predict new phenomena, for example, those occurring in experiments never tried before. If the result agrees with the solution of the problem, this reinforces a theory (though it does not prove it), especially if the experimental result could not be expected without the use of the theory.[46]

This somewhat dramatic kind of reinforcement of a theory does not occur as frequently as one might expect. The most obvious example during the Copernican era is Pascal's prediction that the barometric pressure depends on height.

4.4. Axioms and the aim of science

In Sec. 4.1 we took axioms to be characteristic of a theory. It is not accidental that the Copernican revolution laid more emphasis on the axiomatic method than was done before, because the Copernicans stressed the "mathematical method" to be used in natural science. They did not always agree about the exact meaning of this term, but most of them identified it with the axiomatic method, the method of investigating nature with the help of theories.

Before, and even after Newton, not many scientists were aware of an important shift in the status of axioms. According to Aristotelians and Cartesians alike, axioms should be evidently true. But Copernicus started from the counter-intuitive assumption that the earth moves, and Newton's view of gravitation contradicted the clear and distinct mechanist idea that the only conceivable action between bodies is action by contact. Newton did not care: "...to us it is enough that gravity does really exist, and acts according to the laws which we have explained, and abundantly serves to account for all the motions of the celestial bodies, and of our sea."[47]

Nowadays, all scientists agree with Newton. Modern science explains the observable from the unobservable.[48]

Aristotle's requirements of axioms

According to Aristotelian philosophy, the axioms, or fundamental premises in any explanation must satisfy three requirements.

1. The axioms must be true, they must be known to be true, and they should be better known than the *explanans*, the state of affairs to be explained.[49] It is logically possible to draw true conclusions from false premises, but Aristotle does not consider this a valid proof. In the middle ages, Thomas Aquinas introduced the dogma that true axioms can never contradict each other. This dogma was especially designed to avoid contradictions between scientific or philosophical, and theological axioms or dogmas. In the 17th century, Leibniz transformed this dogma into a "pre-established harmony".

2. The axioms must be unprovable. This seems to be a matter of definition, a provable statement being a theorem or proposition, not an axiom. However, whereas in the modern view provability depends on the context, Aristotle intended it to have an absolute sense.

3. The axioms must be evident. It must be intuitively clear that axioms are necessarily true. Up till the 19th century, Euclid's axioms of geometry served as a paradigm in this respect, but since the 19th-century development of non-Euclidean geometries, this requirement is less evident. Descartes and his disciples accepted it, saying that axioms should be "clear and distinct."

Critique

It is no longer believed that axioms must be absolutely true. It is sufficient to accept their truth within the context of the theory. Outside that context, axioms should have a certain measure of plausibility, or approximate validity. In geometrical or ray optics, for example, one axiom states that light propagates rectilinearly in a homogeneous medium. We know this to be only approximately true, and we know that geometrical optics cannot solve all problems in optics. But a sufficiently large number of problems exists which can be solved by geometrical optics, which is therefore a satisfactory and useful theory.

Next, our distinction between provable theorems and unprovable axioms is less absolute than it was for Aristotle. Sometimes, in a theory the roles of an axiom and a theorem can be reversed. Moreover, the axiomatic status of a statement depends on its context. An axiom in one theory may be a theorem in another one. Consider again the axiom of geometrical optics. In the context of wave optics, the statement that light moves rectilinearly is a provable theorem, at least it is proved to be approximately valid. Hence, assuming the theory of waves, we can establish to what extent geometrical optics is true, *i.e.*, to what extent its solutions of problems are satisfactory.

Aristotle's third requirement has changed most drastically. Modern axioms of scientific theories are no longer clear, self-evident, intuitively understandable. The abandonment of this requirement started with the Copernican revolution. Copernicus' view that the earth moves is contrary to the experience of any human being. The law of inertia, stating that bodies move uniformly and rectilinearly if not subject to any unbalanced force is counter-intuitive. Galileo's axiom saying that all bodies fall equally fast in a vacuum is contradicted by our daily experience. Newton's identification of the force of gravity with the force governing celestial motion, and the inherent idea of action at a distance, are by no means clear and self-evident.

Up till Copernicus, uniform circular motion was considered a clear and rational axiom for the explanation of celestial motion. Kepler, Galileo, Descartes and others accepted it, but ultimately Kepler had to admit that the planets move neither uniformly nor in circular orbits. Kepler's discovery had an importance which only became clear long afterwards. He accepted non-uniform elliptic motion on the basis of observations, which he allotted higher priority than immediate, clear, evident insight. Before Kepler, hardly anybody had done such a thing, and afterwards, even Galileo and Descartes refused to follow him.

Descartes especially held on to the requirement that axioms be self-evident. The main difference from Aristotle concerns the basis of self-evidence, the nature of "clear and distinct" ideas. Descartes rejected Aristotle's form-matter scheme of explanation, his theory of four causes, of the four elements, and the distinction between terrestrial and celestial physics. Descartes was a mechanist, even the main philosopher of mechanism. For him, a clear and convincing explanation can only be given on the basis of ideas and concepts derived from mechanics. In a somewhat less dogmatic sense, the same applies to Galileo, Beeckman, Huygens, and Leibniz.

But Newton joined Copernicus and Kepler. Newton did not require the axioms or laws to be clear and distinct. He argued in favour of the universality of gravity on the basis of observation and experiment, and he put forward the thesis that the mechanical principles of extension, hardness, impenetrability, *etc.*, are found from experience as well.[50]

It has taken some time before scientists accepted Kepler's and Newton's view that axioms may be counter-intuitive. The Aristotelian and Cartesian appeal to common sense is too strong. Even Newton's disciples were tempted to demonstrate Newton's laws of motion and gravity to be self-evident.[51] Especially during the second half of the 19th century, when mechanism reached its acme, an explanation was only accepted if it started from clear, mechanical principles. Adherents to the mechanist world view tried to reduce all physical (and non-physical) laws to mechanics. This was even the case with Maxwell, who, however, inevitably arrived at the conclusion that his laws of electrodynamics could not be reduced to mechanical ones. The succeeding development of electrodynamics, relativity, and quantum mechanics has shown that the axioms of physics cannot be grasped intuitively, not being clear and self-evident.

Modern requirements
Nowadays, the criteria for axioms no longer apply to axioms separately, but rather to systems of axioms, to theories.[52]

The first requirement is that axioms must not contradict each other, because this would allow the proof of a statement and, simultaneously, its negation. In fact, this requirement we have seen applies to all statements of the theory. But with respect to data and presuppositions, it is less severe, because these statements are in principle exchangeable.

The second requirement is that the axioms be mutually independent. If one axiom could be derived from the others, it would be a theorem.

The third requirement is controversial. It states that a theory should be complete, *i.e.*, any theorem which can be formulated within the context of the theory must be provable to be true or false. This requirement is mostly discussed in the philosophy of mathematics, especially since K. Gödel in 1931 proved that a theory in which the natural numbers play a part can never be both complete and free of contradictions. For the philosophy of science this controversy is of little importance, because scientific theories are never complete in the above sense.

The aim of science

Connected to the modern view of axioms is the idea of science. The aim of modern science is not to explain phenomena by deduction from immediately clear axioms. This is the Aristotelian or Cartesian ideal of science. Its aim is rather to find the hidden, unknown laws of nature, which determine the phenomena. Since Kepler, science aims to investigate the lawfulness of nature which manifests itself in the phenomena.

This is by no means a generally accepted view of science. We have already seen Kuhn's view of "normal science" as a problem-solving activity. It is also very common to identify "science" with "theoretical thought." It cannot be denied that in science, theories are invented, developed, and used in order to solve problems. Nevertheless there is some point in distinguishing scientific activity from the use of theories, if only because theories are also applied outside science.

Theories have a *definite logical structure*, and logical functions, as we have seen. Theories as such do not have a specific purpose, let alone a logical purpose. As versatile instruments of thought, they can be used for any purpose, scientific and non-scientific. The theories developed in science can be used to solve practical problems, but this we do not consider the original aim of science.

Contrary to theories, science has a *definite aim*, and is characterized by its aim. This aim is the discovery and opening up of natural laws, formulating them as axioms to be used in theories, and connecting them in order to investigate their consequences.[53]

In this view, science is not a set of statements or theories, an amount of knowledge, but an activity, something people do.[54] Theoretical thought, though very important, is only an ingredient of this activity. It also includes observing, experimenting, dissecting plants or animals, designing instruments, preparing samples, calculating, excavating, inquiring, exploring the literature, and so on.

In this conception, scientific activity is directed to the exploration of the *laws* of nature. But laws cannot be directly experienced, they cannot be observed. The function of laws is to be *valid* for individual things, states, and events. The only way to find laws empirically is to investigate particulars. Particular things or events are considered as a *sample*, as *exemplars*, and are not studied as an aim in itself. The samples, if suitably (*i.e.*, randomly) chosen, are supposed to be representative, hence reproducible. Sometimes, the samples are highly idealized — frictionless motion, rigid bodies, chemically pure substances, a space kept at constant temperature. Idealized samples are studied in order to find laws, to be applied in more concrete, hence more complicated circumstances.

This way to do science was probably born during the Copernican revolution, first, because only then people became aware of the idea of natural laws to be discovered by investigation of nature instead of Aristotle's books (see Sec. 12.2); next, because the method of taking idealized samples is contrary to the organistic world view of Aristotelianism (see Sec. 5.3); finally, because it made science intrinsically historical.

The medieval scholars accepted Aristotelian and other ancient insights to be final and complete. Since the Copernican revolution, science is identified with progress, with discovery, with the development and articulation of laws.[55] Science is history, as we shall see in the next section.

4.5. Crisis and revolution

According to Kuhn's *The Structure of Scientific Revolutions*, a period of normal science ends in a crisis, induced by a persistent anomaly or an increasing number of anomalies, problems which cannot be solved according to the accepted paradigm. Eventually, a new paradigm replaces the old one, and this constitutes a "scientific revolution." This view easily leads to a proliferation of "revolutions" and a devaluation of the meaning of this word. In this section we offer some arguments against Kuhn's theory.

The incredibility of Copernicus' theory
For more than one reason Copernicus' theory was incredible, at least during the 16th century. One of the reasons was that it contradicted the generally accepted Aristotelian philosophy, including its physics, cosmology, and astronomy. For most philosophers, the slight advantage of the Copernican system solving one or two astronomical problems did not offset the loss of a coherent system.[56] Perhaps this motivated the opposition against the heliostatic model by Luther, Melanchton and other German protestants (see Sec. 8.3) even more than the supposed conflict with the Bible, which only became significant when Giordano Bruno used

the Copernican theory as propaganda for his heretical ideas.

Bruno was imprisoned in 1592 by the Inquisition, and burned at the stake in 1600, though not because of his Copernican views. Nevertheless, since this event Copernicanism became suspect in the eyes of conservative Roman Catholic theologians.[57] Copernicus' book was placed on the Index of prohibited books between 1616 and 1621, and from 1633 to 1822. Both periods started with events involving Galileo (see Sec. 8.3).

Sixteenth-century astronomers recognized Copernicus' ability as a mathematician. They chiefly admired his "trepidation theory", explaining an effect that Tycho Brahe proved to be spurious.[58] But apart from Maestlin and Kepler, before 1600 no professional astronomer became a Copernican.

The acceptance of Copernicanism and the emancipation of science
Initially, the adherents of Copernicanism only included people who had reasons independent of astronomy to doubt Aristotelian science: Ramus, Stevin, Gilbert, Benedetti, Galileo, Beeckman, Descartes.[59] They considered Copernicus a welcome ally in the struggle against the supreme Aristotelian philosophy. They became Copernicans, not on the force of astronomical arguments, but because Copernicus' theory supplied them with an argument against Aristotelian cosmology. But it is not sufficient to show that a system like Aristotle's has faults. It is only possible to combat such a system if there is an alternative such as was supplied by Descartes' mechanical philosophy (see Sec. 3.4).

Copernicus, Kepler and Galileo did not have a new alternative at their disposal. They had to take recourse to an older alternative, Platonism, Pythagoreanism, or both. This move was not very effective, because in the course of centuries the superiority of Aristotle over Plato had been firmly established. Plato's views do not constitute a firm building like the Aristotelian one. Still, the renewed interest in Plato during the Renaissance must be viewed in the light of the growing resistance against Aristotelianism.

The historical significance of Descartes is that instead of appealing to an alternative ancient philosophy he designed a new philosophy, the philosophy of mechanism, although he was probably not the first mechanist. Descartes is generally considered to be the founder of modern philosophy. In the long run, the contents of his philosophy have not been very influential in physics. The typical Cartesian views on space, matter and motion have retarded more than stimulated the development of physical theories. But the bare existence of an alternative for the previously superior Aristotelian philosophy worked as a liberating force.

If alternative, competitive philosophies are available, each claiming exclusivity and universality, a normal scientist can afford to ignore them both. During the 17th century the separation between science and philosophy made a start. This separation was completed only after *c*.1800, *i.e.*, after the work of the German philosopher Immanuel Kant. The emancipation of science from philosophy was strongly furthered both by Descartes and by Kant, contrary to their intentions.

But ultimately, it was not Cartesianism that became the philosophy of Copernicanism. Not Descartes, but Newton wrought the synthesis of all anti-Aristotelian currents (see Sec. 5.1). This synthesis has philosophical aspects and backgrounds, yet has not a philosophical but a scientific character. The Newtonian synthesis confirmed the emancipation of science from philosophy. Since Newton, the credibility of a scientific theory is no longer determined by philosophical arguments, but by its agreement with other scientific results.

Crisis

We now arrive at an important point of difference with Kuhn's views. According to Kuhn a crisis arises in a field of science if scientists working within the context of a certain paradigm are confronted with an increasing number of anomalies, of problems not to be solved according to the methods of the paradigm.[60]

For example, Kuhn points to the crisis preceding the publication of Copernicus' *Revolutionibus* (1543).[61] Unfortunately, this example does not tally with the historical facts.[62] Before Copernicus the Ptolemaic theory was considered quite satisfactory by all experts.[63] Copernicus himself was the first to signalize a situation of crisis, but he was hardly unbiased. He had an obvious interest in putting the old theory in an unfavourable light. In the introduction to his *Tabulae Prutenicae* (1551), based on Copernicus' calculations, Erasmus Reinhold stated: "The science of celestial motions was nearly in ruins; the studies and works of (Copernicus) has restored it", nevertheless he was not a Copernican.[64] His updated Prussian tables were better than the outdated Alfonsine tables (13th century), but this was hardly due to the introduction of a heliostatic model, and Tycho Brahe found them both unsatisfactory.

Also the publication of Newton's *Principia* was not preceded by a crisis. Except for the conservatives, who held fast to Aristotelian physics, most educated people considered Cartesian physics satisfactory, promising, and acceptable. The criticism of Cartesian physics was primarily leveled by Newton himself, who again had an interest in putting his competitor in an unfavourable light.

Contrary to Kuhn, we argue that a crisis is often an *effect* of the introduction of a new fundamental theory, rather than its *cause*, because a new theory in general contradicts the accepted *presuppositions*.[65] A new theory makes it necessary to adapt the presuppositions. This evokes resistance, and as long as the presuppositions are not adjusted, their adherents are in conflict with those of the new theory.

This is obviously the case with Copernicus' theory, contradicting the most important presuppositions of his time. Hence, the initial response to Copernicus' theory was negative. The first to accept Copernicanism were already doubting the Aristotelian presuppositions, but the crisis became a fact only when Galileo's astronomical discoveries (1609-10) made the new theory a serious threat to Aristotelian philosophy. The great debate concerning the merits of the Ptolemaic system did not take place before 1543, as Kuhn wants to make us believe, but between 1610 and 1640.

Similarly, Kepler caused a crisis within Copernicanism, by dropping the idea of uniform circular motion in favour of non-uniform elliptic motion. Also his results were ignored initially, but when accepted, they gave occasion to a fundamental change of the presuppositions of planetary theory, for example, the introduction of the concept of a force as a cause of change of motion.

Newton's theory of gravitation was not the effect of a crisis, but its cause. This crisis did not occur in the theory of gravitation, but in its presuppositions, mechanics and mathematics. In mechanics, the principle of action by contact in a plenum had to be replaced by action at a distance in a void. Newton's theory led to the introduction of integral and differential calculus, causing a crisis in mathematics. In order to avoid this crisis, Newton presented the proofs of his theorems in the *Principia* in a geometric way. Mathematicians have struggled with the foundations of the calculus until the 19th century.[66]

The uniqueness of the Copernican revolution
Without exaggeration, it may be said that Copernicanism caused so many crises that once the battle was over, nearly every presupposition was drastically changed. We call it a revolution, because it resulted in the abandonment of virtually all preceding pre-

suppositions.[67] It was to overthrow the entire scientific and philosophical establishment. Hardly anything of Aristotelian physics survived the Copernican revolution, and about the only reason to discuss Aristotelian physics is to understand the rise of modern science.

In the middle ages scientific activity had an overwhelmingly logical character. Independent research was rarely executed, and the scholars restricted themselves to a logical analysis of the ancient arguments. The prevalent impression was that everything worthwhile to know about nature could be found in the tradition. It is an essential part of the Renaissance spirit to become self-reliant, to start explorations independent of the tradition. The adventurous voyages of discovery in the 15th and 16th centuries have contributed much to undermine the trust in ancient views. Many unheard-of things were discovered, and many time-honoured insights turned out to be wrong. Hence, people became more and more critical of medieval philosophy and science.

From their writings it becomes clear that the Renaissance scientists were well aware of the novelty of their works. The word "new" (*nova*) became a platitude — see Tycho Brahe's *De Stella Nova* (1573), Gilbert's *De Magnete...Nova Physiologia* (1600), Kepler's *Astronomia Nova* (1609), Bacon's *Novum Organum* (1620), Galileo's *Discourses on Two New Sciences* (1638). Descartes wanted to make a fresh start in science and philosophy. Newton set out to develop the results of his predecessors. All this shows a new awareness, the awareness that science is a historical endeavour, a force in history.

Since the Copernican revolution, science is no longer static, but dynamic. It is directed to a continuous and progressive opening up of the lawfulness of reality. In physics and astronomy, neither before nor afterwards did a comparable break with the past occur. Even relativity theory and quantum theory are built on the results of the past, and are in part its consequences. The architects of 20th-century physics always stressed the continuity with 19th-century "classical" physics. In this sense, the Copernican revolution is unique in physics.

The solution of problems is an important means to articulate theories, hence to open up the law side of reality. The Copernicans were fond of problems. Whereas the medieval scholars were mostly interested in *logical* problems, the Copernicans set out to solve *scientific* ones. Problem solving became the life-blood of science.

5. The systematization of knowledge

5.1. The reduction of theories

Until now we were concerned with the structure and functioning of a single theory. We considered a theory to be a deductive scheme, connected with concepts, statements, arguments, and problems. We discussed a number of functions of theories — to describe, to predict, to explain, to solve problems. We paid attention to various functions of statements within a theory — axioms, hypotheses, theorems, presuppositions, data, and problem solutions. Presently we shall discuss another function of theories, acquiring, assessing, and systematizing knowledge.

In this chapter, we shall deal with the relations between various theories. Theories are linked up via data and presuppositions. To find data for a theory, one usually relies on other theories. Most presuppositions of a theory are theories themselves. Problems for a theory are often provided by other theories, in need of the solutions as data or presuppositions. For Newton's theory of gravity, for instance, mathematics and mechanics are presupposed theories, delivering lemma's when necessary, whereas optics is applied to find data.

It seems obvious that theories can only be linked if they share one or more statements. This, however, is hardly ever the case, at least in a literal sense. Usually, a concept or a statement in one theory shows some *similarity* with a statement in another theory. In applied geometry, for instance, it is assumed that a light ray resembles a straight line. The laws according to which light is propagated are similar to geometrical laws concerning straight lines.

Hence, theories can be related only as far as such similarities are recognizable, perceivable. The network of theories is constituted by a logical kind of sensory activity. Therefore, this chapter concerns the sensory aspect of human experience.

In the present section we discuss the first kind of relation between theories. In this kind, the axioms of one theory are "derived" in another one, *i.e.*, the axioms of the first (usually the oldest) theory are perceived to have a similarity with theorems in the second theory.

The Newtonian synthesis

We speak of the Newtonian synthesis[1] because in Newton's work a number of earlier theories were incorporated, notwithstanding the fact that Newton explicitly rejected Cartesian physics. The Newtonian synthesis concerns in particular Kepler's planetary laws and Galileo's theory of projectile motion, but also includes the principle of inertia, the concepts of force and acceleration, the theories of impact by Huygens, Wallis, and Wren, the theory of the pendulum by Galileo and Huygens, the principles of relativity, conservation, and composition of motion, Huygens' theory of circular motion, Torricelli's and Pascal's views on the void, and Boyle's theory of matter.

The synthesis is expressed in the theory of mechanics, treated by Newton in the introduction to his *Principia* (see Sec. 3.6), and in the theory of gravity, constituting the main topic of the *Principia* (1687). Besides, Newton wrought a synthesis of 17th-century optics, by means of his influential *Opticks* (1704).[2]

Newton's unifying work may be considered both the completion of the Copernican revolution and the beginning of a new era in the history of physics, Newtonianism (*c*.1700-1850).

Logic versus physics

Some philosophers[3] tend to doubt this synthesis, in particular the so-called "reduction" of Kepler's and Galileo's laws to Newton's. They argue that Newton's theory is incompatible with the former laws, pointing out that the theory of gravity contradicts Kepler's and Galileo's laws on several accounts. Three examples are frequently given.

1. Kepler's first law states that the planets move in elliptical orbits, with the sun at a focus. The philosopher observes that Newton's theory proves this law false, because each planet's orbit is disturbed by the other planets. A physicist or astronomer counters that this disturbance is small, and unobservable by Kepler's means. Within the limits of accuracy achievable in Kepler's time, Newton's theory proves Kepler's law to be true.

2. Kepler's third law relates the dimension R of a planet's orbit with its period T, by stating R^3/T^2 to be the same for all planets. The philosopher observes that Newton's theory proves $R^3/T^2 =$

$K(M + m)$. In this formula, K is a constant, M is the mass of the sun, and m is the mass of the planet. Because m has different values for the various planets, R^3/T^2 is not the same for all planets. A physicist rejoins that M is much larger than m, such that m, the planet's mass, can be neglected with respect to M, the sun's mass. Within the accuracy of the observations available to Kepler, his law is proved to be true in Newton's theory. The same applies to Jupiter's and Saturn's moons, albeit that M, now the mass of the central body, differs in the three cases. Hence, the constant value of R^3/T^2 also differs, which was known before Newton found his theory. Hence, Kepler's harmonic law is a reasonable approximation in Newton's theory.

3. Galileo's law of fall states that in a vacuum all bodies fall with the same acceleration. The philosopher says that Newton's theory shows the acceleration to depend on the height of the falling body above the earth's surface, as well as on its position (the latitude) on the earth.[4] A physicist replies that Galileo's law concerns laboratory experiments of balls rolling down an inclined plane, for instance, or of pendulums with various bobs. According to Newton's theory, the acceleration of free fall in a space confined to a laboratory is constant within the accuracy of measurements possible during the 17th century. Hence Newton's theory confirms Galileo's law.[5]

The difference between the philosopher's and the physicist's attitude appears to be this. Whereas the philosopher is only interested in the *logical* relations between the two theories, the physicist also takes into account how the laws are justified, finding a large amount of agreement. From a strictly logical point of view the philosopher may be right. But the theories discussed are not logical, but physical. Since the middle ages, physicists are not first of all interested in logical aspects of theories, but in their physical relevance.

Physics versus logic
Although all 20th-century philosophers pretend to be empiricists, emphasizing that scientific results are tentative, stressing that hypothetical law statements cannot do more than approximate reality, they immediately tend to forget these cautions as soon as they start a logical discussion. The arguments of our imaginary physicist derive their strength from the provisional and approximate character of any law statement. Also Newton's law of gravity has turned out to be no more than an approximation, and philosophers now call it false.

The physicist puts forward that Newton's results are not contrary to those of Kepler and Galileo, but show sufficient similarity to explain, deepen, and relativize them.

1. Newton *explained* Kepler's and Galileo's laws by demonstrating these to be close approximations of special cases of his more general theory.

2. Newton's theory *relativized* Kepler's and Galileo's, because it showed their limited applicability.

3. Newton's theory also *deepened* Kepler's and Galileo's laws, first because it connected them, showing that the apparently widely differing motions of a projectile and of the moon are closely related. Next, it demonstrated how Galileo's and Kepler's laws can be extended. It showed, for instance, that the orbit of a planet or a comet may be a circle, an ellipse, a parabola, or a hyperbola. Newton's theory enables us to calculate the force of gravity at the surface of Jupiter or Saturn, and the variation of the acceleration of free falling bodies as a function of the latitude, or of height above the earth's surface.

The deviations from Kepler's and Galileo's laws were largely unknown before Newton wrote his *Principia*. On the one hand, Newton was able to start from the assumed truth of Kepler's and Galileo's laws, on the other hand he was able to indicate how these laws should be corrected.[6]

The principle of correspondence

We should prefer not to say that Kepler's or Galileo's theories are absorbed by, or can be derived from, Newton's theory. We had better say that the axioms of the older theories are justified by the new one, and are valid within a certain margin of accuracy. This means that the old theories, being discredited according to the philosophers, in the physicists' eyes gain in credibility, rather than losing it. It also means that the new theory explains why the old theory could be successful, even if its axioms have only limited validity. Hence it is not necessary to reconsider all problems solved with the old theory.

The relation between a new and an old theory is not always of this kind. Newton's theory of gravitation collides with Descartes' vortex theory, because Descartes' axioms (for instance, the identification of matter and space) deviate too much from Newton's axioms and theorems. As a consequence, all problems solved by Descartes' theory have to be solved again.

Accepting Newton's theory allowed people to continue teaching and using Galileo's and Kepler's laws. But without coming into conflict, it would be impossible to teach Newtonian and Cartesian theories simultaneously.

In the 20th century, the Danish physicist Niels Bohr introduced the *Principle of Correspondence* to describe the relation between successive, non-contradicting theories. He applied it to the relation between quantum and classical physics. It has two aspects. First, it says that every new theory has to explain why the old theories were able to give more or less correct solutions to their problems. Second, in order to find new theories, one may take one's lead from the old ones. In both respects, the principle of correspondence applies to the relation between Newton's theory and those of Kepler and Galileo.

5.2. Putting theories to test

As a second example of how theories are connected with other theories or statements, we shall consider how theories are tested. According to the logical-empiricists, theories are verified by comparing their results with observations. Popper advocates that one should try to falsify rather than verify theories. If a theory is refuted, it should be abandoned, and one should never attempt to save a theory.

Observational tests
A theory can be tested by comparing its results (in particular, solutions of problems) with statements acquired in an independent way. Repeated and independent successful tests do not prove a theory, but reinforce its reliability.

For example, we may compare the calculated position of a planet at a certain time with an observation. It would be very naive to think that such an "observation" is purely sensory. As a matter of fact, an observation is more often than not the result of a theory — for instance, the combined theories of optics and geometry. In observational astronomy, one has to know how a telescope functions, and how the light rays incident from a star are refracted by the atmosphere. Even the barest sensorily obtained information has to be transformed into a statement before it can function in a theory. Hence we do not compare logically achieved theoretical results with sensorily achieved observations. We compare statements, independently achieved by means of different theories.

Such a comparison is not based on an equality, but rather on a similarity.[7] In Newton's theory of gravitation a planet is a body having mass, position and velocity in space, but the theory does not say anything at all about the planet's visibility. In observational astronomy, on the other hand, a planet is an extended spot of light having a certain position and velocity on the celestial sphere. In this theoretical context, the concept of mass does not occur (we cannot observe a planet's mass), and in the two theories to be compared, the concepts of position and velocity are different. If we want to compare a calculation based on Newton's theory with a result of observational astronomy, we must first identify the objects they are dealing with as the same planet, and next we must find corresponding properties — the position and velocity of the planet, either in space or projected on the celestial sphere.

Hence we are in need of a certain measure of correspondence between the two theories in order to compare them. In philosophical discussions similarities are taken for granted, but a physicist has to take care of them.

Testing a theory is not restricted to the comparison of a theoretical result with an observational one.[8] It is just as common to compare the results of one theory with those of a competing one. If the comparison is not satisfying, a weaker theory has to give way to a stronger one. A stronger theory may be one that is better established in various other tests. For this reason, Tycho preferred his modified Ptolemaic system above the Copernican one, because his model was anchored in Aristotelian physics. But the strength of a theory may also be judged on account of its problem-generating capacity (Sec. 4.3). For this reason, Galileo preferred the Copernican system.

The subjectivity of a test

The empiricist view that the testing of a theory is a completely objective affair[9] may be contested. The value of a test is determined by the participants in a debate about the acceptibility of a theory. They have to agree about the relevance of a test, and therefore a test has a subjective as well as an objective side.

The confirmation of a theoretically derived statement is hardly ever the result of its complete equivalence with an independently achieved statement. The statements to be compared contain idealizations as well as inaccuracies. The participants in a logical dispute ought to be aware of this fact, which carries weight in every argument.

When stating that Newton refuted Kepler's and Galileo's laws (see Sec. 5.1), Duhem, Popper and other deductivists assumed that Kepler's and Galileo's empirical generalizations were pretended to be exactly true. Newton did not share this assumption, but taking into account their margins of accuracy, he recognized that Kepler's and Galileo's laws agreed sufficiently with some "exact" theorems in his own theory. Newton's *Principia* contains observational and experimental evidence for both Kepler's and Galileo's laws.[10]

Rigorously holding to the deductivist view would render the testing of a theory impossible, for every observational result is both idealizing and inaccurate.

Falsification

According to Popper it is logically impossible to verify a universal statement. Whereas the number of verifying instances is necessarily finite, the number of cases covered by a general statement is indefinite or infinite, and most of these cases are unavailable, *e.g.*, belonging to the past or the future. In contrast, a general statement can logically be refuted even by a single conflicting instance. In Popper's opinion one should not attempt to verify a conjecture, but to falsify it, or to refute it. Popper calls this the method of *Conjectures and Refutations*, trial and error, learning from one's mistakes.[11]

Scientists should strive to formulate vulnerable theories, as much as possible open to falsification (see Sec. 2.3). If many attempts to falsify a theory do not succeed, it acquires a large degree of reliability. If a theory is falsified, it is done with, and one has to start again with a new conjecture. Popper rejects "conventionalist stratagems", *i.e.*, attempts to save a theory by adapting it, or by introducing *ad-hoc* hypotheses; an *ad-hoc* hypothesis being any hypothesis introduced to save a theory without being able to achieve anything else.

Refutation

The view that falsification is more important than verification is called "falsificationism." Its most naive version says that a law statement can be refuted by even a single counter-instance.[12] If this were true, probably all empirical generalizations must be considered refuted, because each set of data contains a number of errors or inaccuracies.

Kepler was content with his first two laws when the calculated orbit of Mars agreed with Tycho Brahe's observations within a margin of one or two minutes of arc. Suppose Kepler had found some observations deviating more than this margin (as was actually the case during a certain phase in his investigation, see Sec. 6.5). Would that have compelled him to reject his empirical generalizations? Certainly not. He would only have been compelled to reject his law statements if he were convinced that the said deviations were not caused by some error, but were forming a *pattern*. An empirical generalization, arrived at by the discovery of some pattern can only be refuted if the counter-instances equally form a pattern. In other words, an empirical generalization can only be refuted by another empirical generalization.[13]

The absolutism of naive falsificationism is induced by the fact that philosophers often discuss statements like "all ravens are black" as typical law statements.[14] However, only in a purely logical sense would the observation of a white raven refute this statement. In a scientific context it is required that observations are in principle repeatable. A single irreproducible observation of a white raven would be declined as an observational error, or as a freak of nature. Only if somebody would discover an entire set, a variety of non-black ravens, the said law statement would be refuted. But to establish the existence of such a variety an empirical generalization is required.

To save a theory
It should be observed that Popper's falsificationism is not that naive.[15] Nevertheless, the above criticism is applicable to his views. He says that the refutation of a generalizing statement requires a "basic statement." Although he does not admit it explicitly, his comments imply that a basic statement has the character of a generalization. Popper stresses that a basic statement must be reproducible. This puts Popper into an awkward position.[16]

On the one hand, he says one should try to falsify a discovered law statement by means of a basic statement. On the other hand, he rejects the use of evading procedures, of prevarications to escape refutation, to save the theory.[17] But his principle of falsification is just as well applicable to basic statements.[18] Put otherwise, if we succeed in finding a basic statement refuting a previously established law statement, we have simultaneously to accept it as a good sport, and to try to falsify it like any other generalizing statement; which means to save the theory.[19]

Popper and Galileo on Copernicus
Popper praises Galileo because he accepted Copernicus' theory as

a "bold conjecture", and Galileo praised Copernicus for the same reason.[20] With respect to Galileo this is understandable, for Galileo considered Copernicus' theory as a challenge. But Popper's praise of Galileo is strange, for Copernicus' theory was refuted, as Tycho Brahe and his Jesuit disciples argued. Copernicus' theory predicted the occurrence of stellar parallax, which was never observed until 1838. Instead of accepting this set-back as a good sport, Copernicus and Galileo tried to evade it by an *ad-hoc* hypothesis — the hypothesis that the fixed stars are far more distant than had ever before been held possible.[21] This *ad-hoc* hypothesis should be a thorn in Popper's flesh.

For the progress of science, it is fortunate that the Copernicans did not adhere to Popper's prescription. "Contemporary empiricists, had they lived in the sixteenth century, would have been the first to scoff out of court the new philosophy of the universe."[22]

Tycho and his disciples were certainly not wrong in rejecting Copernicus' *ad-hoc* hypothesis. They were careful and conservative but highly competent scientists. Like any human enterprise, science is in need of many types of people, careful besides adventurous, conservative besides progressive, normal scientists besides paradigm builders. Science would be impoverished if all scientists would adopt Popper's conservative philosophy with respect to *ad-hoc* hypotheses.

The consequence of a negative test result

Let us now consider the case of a theory being refuted. In such a case a competent user of a theory considers the solution of a problem unsatisfactory because it does not sufficiently agree with an independently achieved result. According to the logicians, this means the end of the theory.

Indeed, this is often the case, especially with brand-new theories. In the daily practice of a scientist many a new theory is formed, consequences are derived and judged unsatisfactory, after which the theory is rejected. Such theories seldom reach the end of the day, and never a publication.

We shall restrict ourselves to theories having a larger life span, because they have proved their capability to solve a number of problems. Suppose such a theory experiences a negative test — does that mean the end of the theory? This is unlikely.[23]

A scientist being confronted with a negative test result begins to mistrust himself, not the established theory. Has he made mistakes? Are the data rightly interpreted? Next, he critically investigates the reliability of the statement (for instance, the observation) refuting the theory. If that does not help, he critically examines the data used to achieve his result. Eventually he makes sure that the applied presuppositions are justified, or applied in the right way. Even at this stage, when he is justified to put the theory into doubt, he will not yet reject it. He may shelve the result as an anomaly (see Sec. 4.2). Or he may introduce an *ad-hoc* hypothesis, to reinterpret his data such that he finds a more satisfactory solution to his initial problem, or to reinterpret the data used to test the theory. This is what Copernicus did in order to explain the non-observability of stellar parallax.

Even if all this does not lead to some satisfactory result, the theory will not be rejected if, as we assumed, the theory is able to solve several other problems. Of course, now the reliability of these older solutions is called into question. Nevertheless scientists will always prefer to have a faulty theory able to solve some problems above having no theory at all.

In other words, a theory able to solve some problems but failing to solve others will only be rejected if an alternative is available.[24] As long as the alternative is wanting, the old theory will be used as best it may.

Usually the transition to an alternative is no easy matter. The alternative theory may be able to solve the problems left by the older one, but in its turn may lead to unsatisfactory solutions of problems already solved by the old one, or it may contradict some generally accepted presuppositions. Copernicus' theory is the outstanding example.

Hence, the transition from an old to a new theory is not a completely objective matter — witness the conflicts due to the transition from the Ptolemaic to the Copernican viewpoint, or from Cartesian to Newtonian physics.

5.3. The sources of data

The most important link between theories is formed by data. We have seen that data function in theories in two ways: (1) at the input, as initial or boundary conditions, needed to derive theorems or to solve problems; and (2) at the output as independent material for the testing of theories, to check whether the solutions to problems are satisfactory.

In order to be able to play this twofold part, data should be independent of the theory. Naive empiricists like Aristotle, and the logical-empiricists, were of the opinion that the data should be independent of any theory at all. We shall investigate this claim.

With or without the help of other theories, data are found in common sense, imagination, observation, measurement, and experiment. These sources of data cannot be strictly separated, but it may turn out fruitful to discuss them separately. (On revelation as a possible source of data, see Sec. 8.3.)

Common sense

A mostly unconscious and unordered process, experience has the effect that data are stored in our memory. This process begins at birth or perhaps even before, and is stimulated by playing, education, and schooling. Sensory impressions play an important part, as well as dreams and previous experience. Playing, education and schooling have a predominantly social character, and therefore a great deal of experience is not individual but communal. For the use of theories it is possible to appeal to communal experience, to "common sense."

Both Aristotle and Descartes used common sense as a reliable and even primary source of knowledge. There is no need to prove common sense, because "everybody knows." It is needed to prove anything deviating from common sense. Aristotle's axioms are based on common sense, so there is no need to prove them. Descartes begins his *Discourse on Method* (1637) with the famous sentence: "Common sense is mankind's most equitably distributed endowment."[25]

Aristotle's philosophy has been characterized as the philosophy of common sense.[26] For instance, from common sense he takes the view that violent motion always needs a force, and that speed is proportional to applied force, and inversely proportional to resistance. It is based on common experience with a horse drawn cart. Also the statement that for natural motion no force is needed is taken from common sense — everybody knows that a stone falls by itself.

To establish the truth of common sense knowledge no specific observation or experiment is needed, nor is such truth founded on sensory experience exclusively. Common sense knowledge is self-evident, it is intuitively understood by the mind, enriched by experience.

In order to make use of self-evident truths, we have to become conscious of them, we have to transform our experience into a statement, because only statements can function in a theory. Plato called this process of increasing awareness "*anamnesis*" (recollection). A dialogue between a teacher like Socrates and a pupil is a useful instrument to awaken the subconsciously present knowledge.[27] In his *Dialogue* (1632), Galileo applied the Socratic method of *anamnesis* repeatedly.[28] Plato's view that intuitive experience is the most important source of knowledge is strongly determined by his philosophy of ideas, which can never be known by observation or experiment (see Secs. 2.1, 3.2, 12.1). Aristotle was less idealistic, and rated observation higher than Plato did.

Since the Copernican revolution, the importance of common sense as a source of data for theories has drastically diminished. In the natural sciences it has virtually disappeared — modern science is counter-intuitive (see Sec. 4.4). This has been a slow process. The Copernicans appealed to common sense if they could utilize it in their polemics against Aristotelians, but they became increasingly aware of the uncritical nature of common sense knowledge. Galileo, for instance, criticized the common sense view that light is propagated instantaneously.[29]

More than before the Copernican era, it is now realized that common sense knowledge is largely determined by cultural and educational contexts. For instance, it is nowadays common sense that the earth rotates daily about its axis, and annually around the sun.

Imagination

Imagination builds on experience and transcends it. By his fantasy man is able to create data out of experience based on sensory impressions. Ideal mathematical figures like circles and triangles, or ideal instruments like the rigid lever or the frictionless horizontal plane, are products of human imagination. Nevertheless, according to Plato and Aristotle, these "ideas" or "forms" have a universal and eternal character.

Imagination is an important aid for any source of data. Direct experience and results of observation and experiment are usually too complicated and too fragmentary to be applicable in a theory. They have to be put into a suitable form, at least the form of a statement. Generally this implies simplification, interpolation, and extrapolation.

Suppose for instance that the measurement of two related variables being recorded graphically leads to the conclusion that the measured points constitute a straight line. On behalf of a theory we assume that the two magnitudes are proportional to each other, tacitly appealing to our experience that deviations from the straight line may be due to measurement errors. By interpolation and extrapolation we imagine that the discovered relation also applies to values we did not measure. The reduction of a finite set of measurements to a statement about the proportionality or any other relation between magnitudes is clearly an act of the imagination.

Also the data used in the numerous problems submitted to students are mostly imaginary. The search for data in a book of tables requires a certain amount of imagination. An experienced researcher knows what to seek, where to look, and how to interpret what is found.

Imagination is a gift, which some people have more than others, and a craft which can be learned and developed. Of course, it is more than a source of data. For instance, imagination is also needed for the development and application of theories.[30]

Observation

We take "observation" to be more than "sensory impression" or even "perception." Observation is directed by one's attention. Hence, data achieved by observation are *expected*, they never come unawares. Observation requires training, and a certain amount of background knowledge.

Empiricists take observations to be "hard facts", independent of any theory. Aristotle was a keen observer, especially as a biologist, and all Aristotelians took direct observation to be a primary source of data. Both 19th-century romantic philosophers and the logical-empiricists shared this view. On the other hand, Plato thought sensory experience to be deceptive. With respect to celestial phenomena, he said: "These can be apprehended only by reason and thought, not by sight..."[31] Under his influence, the Copernicans became aware of the need to be critical with respect to direct observation.[32]

First, they discovered that observation can be enormously enhanced by the use of instruments like the telescope and the microscope. Secondly, they found it necessary to *theorize* observation, *i.e.*, to criticize observational results with the help of theories. Thirdly, they depreciated sensory phenomena to "secondary properties", to give way to the "primary" mechanical or atomic properties of matter (see Sec. 3.4). They restricted observation to its physical aspect, the interaction of the investigated system with (preferably) an instrument.

To modern science, observation is theoretically directed to a *segment* of reality.[33] "Reality" is a theoretical concept. It is an extension of the experienced environment which is bounded up by the horizon of experience, both individually and communally. The experiential horizon widely differs between men and even between communities. The concept of "reality" implies the view that the experiential horizon of common sense is not the end of all possible experience.

Attentive experience intends to enlarge the experiential horizon. This process is tremendously advanced if the attention directing observation achieves a theoretical character. In that case, it is a theory which conducts the act of observation.

Obviously, we have to pay a price for this advance. Because observation is directed to a segment of reality, usually outside our ordinary experiential horizon, we lose sight of the rest of reality. The directional observation leads to a fragmentation of reality. This becomes even more apparent if our sensory observation is extended by the use of instruments.

Ancient and medieval philosophy were strongly concerned with the totality of our experience. In Plato's and Aristotle's philosophy, theoretically directed observation did not play an important part. However, it did so in several special sciences, developing as autonomous parts of philosophy, such as astronomy and optics, and also in biology, in which Aristotle obtained a classification of animal species based on observation.

In particular the use of instruments to enhance observation is in need of a theoretical foundation. Designing instruments, using them, and expressing observational results into observation statements, all require understanding of the functioning of the instruments on the basis of experience, imagination, and theories.

If such a theory is not available, the observations acquired with an instrument are unreliable and incredible. Therefore it is not strange that Galileo's compatriots initially rejected his observations of Jupiter's moons. In 1609-10 no theory was available to explain the functioning of a telescope. Therefore, some people refused to look through Galileo's telescope, or declared Jupiter's moons to be produced by the telescope, like some other effects such as discolourment and double vision.[34] It was later, in his *Dioptrice* (1611), that Kepler was able to lay the foundation of an explanation of the telescope.

The significance of directed and instrumental observation for the collection of data has ever increased. There is an intimate connection between the formation of theories and observation. The collection of data takes place within a theoretical context, determining which data are relevant, how they are obtained, corrected, and digested.[35]

Measurement

A specific way of observing is measuring, the establishment of numerical relations by means of a scale.[36] Here, instruments are indispensable — a metre stick, a balance, a sextant, *etc.* Because Aristotelian science was mostly qualitative, measurement never played a part in it, except in astronomy, considered to be part of mathematics.

During the Copernican revolution the emphasis shifted to a quantitative view of science. At first this was attempted in the Pythagorean way, especially by Kepler (Sec. 2.5). Gradually, it became clear that physics is not a science of pure numbers or spatial relations, but of *magnitudes* like mass, quantity of motion, or temperature.

Instrumental measurements led to a critique of common sense perception, for instance in the case of the rectilinear propagation of light. Since the 16th century astronomers became aware that light is refracted by the atmosphere, but still at the end of the 17th century, Newton had to teach the astronomer Flamsteed how to correct measured star positions for this phenomenon.[37]

During the Copernican revolution we see a growing awareness of observational errors, and of the limited accuracy of measurements.[38] This started with Tycho Brahe, who not only constructed much better instruments than anyone before, but also added tables with corrections to each instrument.

Experiments

Directed observation widens the horizon of experience, but leaves reality more or less untouched. But an experiment interferes with reality, and changes it.[39] Even more than a directed and instrumental observation, an experiment gives rise to isolation and idealization. In an experiment a part of reality is isolated in order to control states, events, and processes, such that they can be kept constant or change in a determined way. By an experiment reality is manipulated.

But the manipulated reality is an idealized one.[40] External influences are excluded as much as possible or considered irrelevant and ignored. The substances involved are not natural but artificially purified — pure chemical substances, or genetically pure plant and animal stocks. Also the analysis of the experiment, the interpretation of the acquired data, is highly idealized. In the report of an experiment a sketch is given of the apparatus, showing only the most relevant details.

In a logical sense, any experiment has a physical character, whether it is performed in physics, chemistry, biology, sociology, or elsewhere. In an experiment, first a "system" is defined, isolated from its environment, as much as possible, and as far as relevant. What is considered relevant must be decided by the experimenter. In a chemical experiment other things are relevant than in a biological or sociological experiment. Next, something in the system is changed, either by some external influence, or by some internal process started by the removal of some prohibition. Finally, the result is observed and recorded. This means that any experiment is determined by some cause-effect relation, and we have seen that the relation of cause and effect is a physical analogy. Such an analogy cannot be found in pure mathematics, which, therefore, is not an experimental science.

Whereas directed observation played a modest part in ancient and medieval science, experiment was even explicitly rejected as a source of knowledge.[41] Admittedly, some medieval scientists like Roger Bacon (c.1270) propagated the performance of experiments, but probably they did not intend to proceed beyond common sense or daily experience. They stressed the need of what they called experiments as a counter-balance to the awe for written works inherited from the ancients.[42] The experiments done by alchemists or physicians were not intended to collect data for the formation of theories, but to pursue practical aims like making gold or medical drugs.

In fact the artificial character of experiments made them a suspicious source of knowledge. Ancient and medieval physics was concerned with the nature, the essence, the character of anything presenting itself to experience. The artificial did not belong to physics but to mechanics, the art of making machines. Therefore, in medieval *arts* experiments could take place, and mechanics came to prosperity, but in *science* experiments were excluded. In science, considered a passive contemplation of nature, only experience, imagination, observation and logical analysis could function as sources of knowledge.

This view changed gradually but radically during the 17th century, when the organistic world view was replaced by a mechanist one, which put man opposite nature. Now it seems fit to use experiments as sources of knowledge. In the 17th century, scientists became interested in experiments more than ever before, seeking contact with instrument makers. But even in the first half of the 17th century, several different views on the significance of experiments can be discerned. We mention three of them.

Francis Bacon was not a scientist but a civil servant and a philosopher. As a pragmatist, he rejected Copernicanism because he did not see the use of it. His books, in which he criticized Aristotelianism, were very influential, especially in England.[43] He propagated the use of experiments and observations, in order to collect data necessary to find laws by means of induction. Although Bacon did not deny that logical deduction plays an important part in science, he is generally considered the herald of inductivism and empiricism. Bacon conceived the data found by observation and experiment to be independent of any theory. He also invented the idea of an *"experimentum crucis"* as a means to decide between two rivalling theories or hypotheses. The result of a crucial experiment should be predicted differently by the two theories.

Also *Galileo* used experiments to investigate the lawfulness of nature. For him experiments have the function of testing law statements and theories. Usually he took his input data from common experience. Experiments are used at the output of his theories.[44] Being a skilled instrument maker himself, more than Bacon, Galileo realized the complexity of experiments. Therefore he sometimes restricted himself to thought experiments, appealing to experience and imagination.[45] For instance, Galileo discussed the question of the comparative height to which a ball will roll upward on an inclined plane, if it has first rolled down another inclined plane, then moved along a horizontal plane. There is no evidence that he ever performed this experiment in real life. Other examples of thought experiments are Galileo's discussion of the motion of pendulums of various lengths, of ballistic motion, and Simon Stevin's discovery of the composition of forces.

Galileo's reservations with respect to experiments are clearly expressed by his comment on his proof that a cannon ball reaches farthest if shot at an inclination of 45°: "...to understand why this happens far outweighs the mere information obtained by the testimony of others or even by repeated experiments...The knowledge of a single fact acquired through a discovery of its causes prepares the mind to understand and ascertain other facts without need of recourse to experiment..."[46]

Finally, *Descartes* valued experiments less than Bacon and Galileo did.[47] He was a rationalist, believing that the most important laws of nature are found by theoretical reasoning alone. A theorem logically deduced from clear and distinct ideas does not require verification. Any conflict with the results of observation or experiment is due to the complexity of reality (see Sec. 3.4).

However, Descartes accepted experiments and observations as sources of input data. He felt able to deduce that the refractive index of light is independent of the angle of incidence, and is determined by the properties of the two substances involved, for instance, air and water. But the actual value of the index of refraction can only be determined in an experiment.

Hence we see that for Bacon experiments have a *heuristic* function, because they give rise to the development of new theories. For Galileo, experiments are mainly useful for the *testing* of theories, and for Descartes, they are helpful to obtain *input data*. Modern philosophers seem to reject Bacon's view, to accept Galileo's, and to ignore Descartes'. Scientists, on the contrary, apply all three of them.

5.4. Objectivity

For any empirical theory, the data obtained by means of observation and experiment are highly relevant. The data functioning in a theory always have the form of statements, hence observations and experiments must be interpreted by observers and experimenters. There is no straightforward, mechanical or automatic relation between theory and observation or experiment.

Test statements have a logical character no less than other statements in a theory. For instance, an observation statement can only function as a datum if it is clear what element in the statement is distinguished from whatever other element, and the statement must be unequivocal.

The objectivity of data
Moreover, the data must have an objective character, especially if intended to test a theory. Some logical-empiricists thought to be able to achieve this aim by assuming that the observational results, the "sense data", are completely independent of any theory,[48] but this view seems to be untenable. Some modern philosophers tend to adhere to the other extreme by assuming that any observational result is completely determined by a theory.[49]

An intermediate position seems to be more in accord with scientific practice. We argued that observational results can only be acquired in a theoretical context, but nevertheless have a certain autonomy, like any statement (Sec. 1.5). They can be transferred from one theory to another one. If this were not the case theories could not be tested, and competing theories could not be compared.

The extreme view that each observational statement is completely determined by its theoretical context would imply that each observational result achieved within the context of a rejected theory would lose its validity. In fact, although the theories of Tycho Brahe and even of earlier astronomers have been abandoned for a long time, their observations are still in use. Nevertheless, each observational result must be critically investigated in view of its theoretical presuppositions, and is liable to be corrected or rejected because of new theoretical insights.

As soon as an observation result is obtained, based on certain theories and experiments, and is published, it can be used by whomever likes to do so. An astronomer may use observation statements about stars he has never seen himself. He may even accept these results if he does not understand how they are found. This possibility gives rise to a division of labour, for instance between experimental and theoretical physicists.

The possibility to test theories with the help of statements acquired by means of other theories is based on the fact that theories form a *network*. In this network each theory has a limited autonomy, warranting the possibility to use data obtained outside its context. The larger this autonomy, the better a theory can be tested independent of the other theories of the network, the "unproblematic background" of the theory.[50] But a theory having no ties at all with other theories can only be tested in a logical way, by looking for internal contradictions.

The objectivity of a measurement
The objectivity of a measurement is considerably larger than the objectivity of an observation lacking a quantitative aspect. The establishment of the temperature by means of a calibrated thermometer allows one to compare temperatures at different times and places such that to a large extent the subjectivity of the observer is excluded. This form of objectivity requires first of all the availability of instruments, next a social organization enabling agreements about scales, in particular.

During the middle ages such an organization was lacking, at least above a municipal or regional level. Standards of measures and weights differed from place to place. The development of a uniform system of units had to wait for the French revolution which introduced the metric system.[51]

Reproducibility
So far we considered objectivity especially in contrast with subjec-

tivity arising from the insight of those who use the theory, make observations, and do experiments. It is impossible to achieve complete objectivity, and the secret of the art is to strike a good balance between subjectivity and objectivity.

Another scientific criterion for objectivity concerns experiments — these have to be reproducible. The results of a physical experiment must be independent of the time when, and the place where the experiment is performed and even the state of motion of the whole experimental set-up.

This form of objectivity rests on a theory called the "principle of relativity." According to this principle, in a closed system physical interaction is independent of time, position and motion. Put differently, the physical mode of experience is irreducible to the numerical, spatial, and kinematic ones (see Sec. 3.6). Hence also the laws determining interaction must be formulated independent of time, motion and position.

This only concerns the system's *external* relations of time, position, and motion. The investigated *internal* interaction may depend on the temporal duration of processes within the system, the relative position of mutually interacting substances or bodies belonging to the system, and on their relative motion. For instance, the gravitational force depends on the mutual distances of the bodies concerned, and the frictional force depends on their relative motion.

Only through Einstein's theory of relativity has the significance of this kind of objectivity become clear, just like its relation with the laws of conservation of energy, linear and angular momentum, and motion of the centre of mass of the system as a whole. The foundation for this development was laid down in the 17th century, however.

In ancient and medieval theories, in which everything was accorded its natural place, relativity was out of the question. But in order to make plausible that the rotation of the earth cannot be experienced, Galileo had to demonstrate that numerous processes occur despite the motion of the system as a whole. Newton still contemplated the existence of an absolute space and time, but Huygens applied the principle of relativity to the problem of impact, and Leibniz was the first to state explicitly the relational character of space and time.

The objective validity of natural laws
One of the chief characteristics of modern science is the assumption that natural laws found by means of theories, observations,

and experiments, are universally valid — valid for everybody, everywhere, and always. They are valid for everybody irrespective of race, prosperity, political or religious conviction. They are even supposed to be valid apart from whether people accept or reject them, and whether they are understood or not.

The supposition that laws are valid everywhere is necessary for astronomical theories. The results found in laboratory experiments can be applied to stars, and *vice versa*. And the supposition that laws are always valid is a starting point of any theory of evolution, whether astronomical, biological, or geological. This so-called "principle of uniformity" was founded during the Copernican revolution by the rejection of any fundamental distinction between terrestrial and celestial physics.

The objective validity of natural laws constitutes the basis of the objectivity obtained by reproducibility. It has induced some people to replace objectivity by "intersubjectivity." A statement is objective if it is shared by several people and preferably by all.[52] This view is mistaken, however. The subjectivity of a statement is not inversely proportional to the number of people who accept it. Of course, intersubjectivity may have its ground in objectivity; the reason why people share certain views may have an objective ground.

The systematization of knowledge

Whereas in the preceding chapters we mostly discussed theories as independent autonomous entities, in this chapter we found that theories can only be fruitful if they are multiply connected. This shows another function of theories, besides prediction, explanation, and problem solving. It is the systematization of knowledge, *i.e.*, to bring pieces of knowledge into contact with each other. Also this function of theories is governed by the principal law of logic, the law of non-contradiction. We can only achieve a systematic, coherent theoretical view of reality if statements contained in one theory do not contradict those of other theories.

This aim cannot be achieved in one stroke, and possibly it will never be achieved to any degree of perfection. It is an ideal of science, to be approximated but never to be reached. As a matter of fact, the more problems we are able to solve, the more unsolved problems we seem to discover. Nevertheless, the pursuit of systematic knowledge is worth the effort.

6. Heuristics

6.1. Induction

In Secs. 4.4 and 4.5 we pointed out the historical character of science. Since the Copernican revolution, science aims at the opening up of the lawfulness of nature.

In Chapter 6 we pay attention to the way theories are discovered, that is, to *heuristic*, the art of discovery.[1] This concerns first of all the question of how universally valid law statements are found. According to Aristotle, universal statements spring from experience, and derive their validity from theoretical thought.[2] Their self-evident truth is grasped intuitively. Francis Bacon sought the source of all knowledge in observation and experiment. By "induction" observational data are generalized into law statements. Descartes required the most general laws to be deducible from clear and distinct ideas.[3]

Popper finds the source of general law statements especially in the human imagination. By putting forward vulnerable bold conjectures, having a large capacity for the solution of problems, science contributes to the growth of knowledge. Popper is very critical of induction.[4]

Pattern recognition
We defined a theory as a deductively ordered set of statements. Induction, being understood as the generalization of a limited number of factual statements, is not deductive, and can neither be theoretical nor theoretically justifiable. In Sec. 4.4 we observed that science is more than theoretical thought only, the aim of science being interpreted as the search for universally valid laws of nature, with theoretical and other means. These laws are hidden, they cannot be "observed", except as far as they turn out to be instantiated. In an empirical way laws can only be found by studying phenomena — things and events showing some pattern. Hence, though not theoretical, induction may very well be scien-

tific, an indispensable procedure of science.

Popper's and other deductivists' radically negative attitude against induction does no justice to scientific practice. For, in spite of all philosophical objections against induction, it cannot be denied that generalization based on a relatively small number of observations is commonly applied in natural science. This kind of induction is based on the recognition of a *pattern*, founded on similarities, combined with previous experience with similar situations.[5]

For instance, if a scientist draws a straight line through a small number of points in a graph and then concludes that there is a proportionality between two measured quantities A and B, he makes use of his insight into similar situations. It is the insight that a correlation between the properties concerned may exist, the recognition that such correlations are often quite simple, that small deviations between the observed points and the straight line can be ascribed to measurement inaccuracies, and so on. Of course, this method is fallible.[6] If a physicist doubts his results, he may do some more experiments, or he may ask a colleague to repeat them. Never will it be possible to achieve absolute certainty. But for practical reasons, a series of experiments will sooner or later be terminated.

According to Popper, the procedure is the other way round. A physicist starts with the conjecture that A and B are proportional, and his measurements are intended to refute this. When he has convinced himself that he cannot falsify his conjecture, he should publish his results inviting other physicists to try to refute them.

So far, the difference between the inductivist's and Popper's accounts seems to be rather subtle. It will be difficult to make a distinction between the two descriptions of what a physicist actually does. But let us proceed with our example.

Suppose a physicist conjectures A and B to be proportional, and he wants to establish the proportionality constant, $c = A/B$. In which way should he proceed according to Popper? Usually, a physicist performs a number of experiments and renders his results graphically in order to determine the value of c from the slope of the line. In particular, the establishment of the values of the so-called constants of nature (like the speed of light) is basically inductive. A Popperian may answer that after the value of c has been found, the generalization "A/B is always equal to c" should be tested. This should be admitted. But the point at issue is that (according to Popper) only this critical stage (starting from the conjecture "A/B is always equals to c") is rational. The preceding inductive stage, the initial determination of the value of c, belongs to the "context of discovery", which is not open to rational analysis, according to Popper.[7]

Kepler and Newton
Popper says that the starting point of any theory is a bold conjecture. For example, Popper points to Kepler's laws, suggesting that Kepler invented these laws, and then put them to the test.[8] This account, however, is contradicted by historical evidence. In *Astronomia Nova*, Kepler describes how he wrestled with his data in order to bring them into accord with Copernicus' theory. After several years of hard labour Kepler had to abandon his attempts. He announced his first two laws, not as bold conjectures, but as the result of a six year struggle with Tycho's data.[9] His results did not have the character of a conjecture, but of a pattern recognition.[10] After extensive calculations Kepler recognized the pattern of non-uniform elliptic motion.

Kepler's model of the planetary system, published in his *Mysterium Cosmographicum* (1597, see Sec. 2.5), connecting the dimensions of the solar system with those of the regular bodies, can be considered a bold conjecture in Popper's sense. Although Kepler soon realized that the model was falsified by the observations, it continued to fascinate him throughout his life.[11] On the other hand, Kepler's third law (the harmonic law) can only be seen as an empirical generalization, based on induction. It is quite impossible to assume that the constancy of R^3/T^2 was found as a "bold conjecture."

Popper also calls Newton's law of gravity a bold conjecture. Here we have a rare case in which Newton himself rejects this view. When Newton was writing the *Principia* and the first draft reached London, Hooke learned that Newton applied the inverse-square law to problems concerning planetary motion. Previously, in a Popperian way, Hooke had conjectured that the celestial bodies attract each other according to an inverse-square law, and he demanded Newton's recognition of his priority. Newton rejected this demand indignantly. According to Newton it deserves no merit to posit a law statement. His merit was to derive the law of gravitation from the phenomena, by mathematical analysis. Hooke was unable to do anything of this kind.[12]

Induction and deduction
We considered Kepler's laws as instances of "empirical generalizations" (see Sec. 2.2). In this respect they differ remarkably from

Ptolemy's and Copernicus' results, as well as from Newton's. Ptolemy and Copernicus started from the axiom of uniform circular motion, not based on observation but on a philosophical argument. Of course both made use of observation results to determine the parameters of their circular motions, which in no way can be found by conjectures or by reasoning.

On the other hand, Newton succeeded in deriving the laws of planetary motion from a law found on the basis of mathematical reasoning (see Sec. 6.4). Hence we see three ways to find the planetary orbits: by hypothesis, by deduction from another law statement, or by the generalization from observational results.

We agree with Popper and other deductivists that a generalization from a limited number of observation results cannot constitute a *proof*. But like Kepler, Galileo and Newton, we assume that the relative truth of an empirical generalization can be established given the truth of the observation statements from which it is derived. Pattern recognition leading to empirical generalization is an important way to distinguish relative truth from relative falsity. As such, induction, the establishment of empirical generalizations, belongs to logic, one of whose functions is to distinguish between true and false statements.

Deductivists deny this. Popper rightly stresses that induction cannot lead to certain knowledge of laws. He emphasizes that no methods exist to arrive at certain knowledge. Hence, also logical deduction, and the method of conjecture and refutation, are fallible.[13] Nevertheless, Popper considers deduction a valid logical procedure (although it is fallible), whereas induction is invalid (because it is fallible). We arrive at a different conclusion. Because induction is fallible, it must continually be criticized with the help of theories — but this thesis can easily be reversed and applies to theoretical deductions as well.

Naive inductivism

Inductivism as a hypostatization of induction may be rejected without abandoning induction as a valid method of science. Though not theoretical, induction is logical, as far as it is a method to distinguish true from false statements. Ignoring statistical methods, Popper states that inductive procedures do not exist.[14]

One of Popper's objections against empirical generalizations concerns the somewhat naive inductivist view that these generalizations can be made without theoretical presuppositions, without prejudice. This objection is fully justified.[15] No scientific investigation into the correlation between a number of magnitudes or phenomena can be started unless the investigator at least conjectures the possible existence of a correlation. Usually he has an idea beforehand about the form and extension of such a correlation. This was also the case with Kepler, who made his discoveries while working alternatively in the context of Ptolemy's, Copernicus', and Tycho's theories.

Nevertheless, empirical generalizations also have a certain autonomy with respect to their theoretical context. In Kepler's case this is very clear, because his first two laws refuted all previously conceived theories. Thereby his laws gained a plausibility independent of any theory of planetary motion, because Kepler did not have an alternative to the theories of Ptolemy, Copernicus or Tycho. This alternative was only found by Newton.

Studying *Astronomia Nova*, one cannot escape the conclusion that Kepler's investigation of the motion of Mars was highly imaginative. It shows that both inductivism, and Popper's criticism of induction, overlook the creative role of the human imagination in the process of induction. Inductivism makes induction an automatism, an infallible tool, applicable in a mechanical way. To find generalizations is a completely objective affair in this view. On the other hand, Popper thinks that generalizing statements are products of the human imagination and creativity, hence purely subjective. Popper's criticism of induction stems from his view that objectivity and subjectivity are contrary and unbridgeable. They belong to different "worlds" (see Sec. 1.2). This view prohibits him from understanding that induction is a process having both objective and subjective aspects. Kepler's laws are based on solid facts (Tycho's measurements), but could never have been found by somebody lacking the creative imagination of Kepler.

6.2. The method of successive approximation

Until recently heuristic was hardly considered a subject of science. Koestler got the impression that the Copernicans worked without method, without knowing where to go, like sleepwalkers.[16] The logical-empiricists and the deductivists sharply distinguished between the "context of discovery" and the "context of justification", only the latter belonging to the realm of philosophy.[17] The context of discovery was the concern of history and psychology, and was not considered to be open to logical analysis.

Lately, philosophers have started to think otherwise. For

example, Kuhn, Hanson, Feyerabend, Lakatos, and Laudan have paid attention to the way theories are discovered.[18] Especially Lakatos introduced heuristic as a philosophical subject, albeit that he restricted it to finding solutions to problems. On the other hand, Feyerabend states that no real scientist ever holds unwaveringly to the canons of philosophical methodology.

Lakatos' methodology of scientific research programmes
Lakatos considered himself a pupil of Popper, although Popper rejects the honour.[19] Trying to arrive at a synthesis between Popper's and Kuhn's views, he developed a new philosophy, called the "methodology of scientific research programmes."[20]

Lakatos' initial intention was to find a criterion for choosing between various research programmes.[21] Scientific research becoming ever more expensive, it would suit governments and other science supporting agencies to have an objective criterion for distributing their money. Later, Lakatos admitted that his method does not fit this purpose, but he suggested that in historical situations it is possible to make clear why a certain research programme deserved preference over another one.

Lakatos' starting point is to compare research programmes rather than theories. A research programme is not a theory, but a series of theories, together with a certain heuristic. The theories in the programme have a number of suppositions in common. This is called the "hard core" of the programme. The adherents of the programme accept the hard core to be true and unassailable. They will beat off any attack on the hard core, any attempt at falsification. For this purpose, they have at their disposal a "negative heuristic", a set of rules saying what has to be done to defend the hard core. The statements, eventually *ad-hoc* hypotheses, needed to achieve this aim form a "protective belt" or a defense system surrounding the hard core.

Besides, the programme has a "positive heuristic", again a set of rules according to which problems can be generated and solved. The hard core and both heuristics form Lakatos' variant of Kuhn's paradigm, with one important difference. Whereas Kuhn suggests that in every field of mature science only one paradigm can be operative, Lakatos produces historical evidence to demonstrate that usually two or more competitive research programmes operate simultaneously in the same field.

In order to find out which programme deserves preference, Lakatos makes distinction between "progressive" and "degenerative" programmes. A research programme is progressive if it predicts or produces new facts. It is "theoretically progressive" if it predicts some novel unexpected fact. It is "empirically progressive" if the prediction turns out to be true.[22] A progressive research programme uses mostly its positive heuristic, it solves its own problems. A research programme is degenerative if it uses mostly its negative heuristic. It achieves no more than the invention of *ad-hoc* solutions of problems generated by a competing programme.

Within a research programme every theory is succeeded by the next one because of a problem shift.[23] It is therefore clear that in Lakatos' view problem solving is the most important part of scientific activity.

Newton's research programme
Lakatos and his disciples have investigated several historical situations, and have written interesting monographs, applying the "methodology of scientific research programmes."[24] Lakatos himself studied Copernicus and Newton, among others. Let us first consider Lakatos' view on Newton's research programme concerning the solar system. Lakatos describes this programme as a series of theories.[25]

The first theory states the sun to be fixed, and studies one pointlike planet moving around the sun. Next, the sun is no longer fixed, but travels together with the planet around their common centre of mass. Then the two bodies are considered to be spheres. In the next theory the planet is a sphere rotating about its axis. Then other planets are introduced, as well as their satellites. Again, a new theory concerns the influence of the satellites on the tides, and so on.

Lakatos considers these successive stages in Newton's programme as different theories, with the law of gravity and the three laws of motion as a hard core, besides the supposition that gravity is the only force operative in the solar system.

In our terminology, we would not consider these stages as a series of theories, but as a series of problems, to be solved by means of one theory (characterized by its axioms), with a varying set of data, or initial conditions. At one stage it is given that the sun and the planets are points, at the other stage that they are spheres. Also Lakatos admits that the various stages are characterized by problems — each new "theory" is needed because of a problem shift. In order to prohibit a profusion of theories it seems preferable not to speak of a succession of theories, but of a succession of "models." This links up with common usage.

Models

Lakatos himself defines a "model" as a "set of initial conditions, which are known to be replaced sooner or later during the development of the research programme."[26] It is even known beforehand in which respect the models will eventually be changed. For instance, when Newton investigated his model of a pointlike planet, he already knew that he would replace this model by that of a spherical planet. The function of a model is not to provide a picture of reality, but to formulate a solvable problem. Therefore a model is an idealization. One starts with a relatively simple problem in order to investigate the difficulties and potentialities of the theory, and to master them. Solving each model step by step, introducing ever more detail, one hopes to find a way to tackle the real problem — in Newton's case, the problem of the solar system.

Indeed, this step by step method is widely applied in the natural sciences, especially in the study of the structure of matter — atomic physics, solid state physics, astrophysics, *etc.*[27] It even has a name — it is called the *method of successive approximation*. It consists of the construction of a series of models, each tentatively and temporarily accepted as an approximation of the real structure, whereas each new model is considered a better approximation than its precursor. If that turns out to be the case, the series might be called "progressive."[28]

Falsification

Lakatos emphasized that there is little sense to try to falsify such a model. It is already false at the moment it is conceived, it is "born refuted." This is only partly true. First, we have seen that within each theory any statement has to be considered true. Therefore, in principle any model can be falsified by proving that it contradicts the presuppositions or axioms of the theory. This was for instance the case with one of Newton's first models, the model of a fixed sun and a moving planet. This model contradicts Newton's third law (of action and reaction), as Lakatos observes.

Secondly, some models will be recognized to approximate empirical generalizations. Newton's model of a moving sun with one planet satisfies Kepler's first two laws. Such an agreement shows that the series of successive models is on the right track.

Thirdly, the idealized data should not deviate too blatantly from accepted facts. Thus, Newton could take the planets to be pointlike when studying their motion around the sun, but he could not maintain this model when considering the motion of a falling body near the earth's surface. He had to make sure that the inverse-square law is also valid if the earth is considered a huge sphere.

Competition

If two competing research programmes are available, Lakatos assumes that at most one of them can be progressive. Hence, in the case of Newton's programme, Lakatos has to show that the competing Cartesian programme was degenerative. This turns out to be quite difficult, for the Cartesians and the Newtonians were usually concerned with different problems. Part of Newton's hard core was his world view of point masses in a vacuum interacting through forces. In the 18th century it was successful in solving many problems in the fields of gravity, planetary motion, and the initial development of electricity and thermal physics. The hard core of Cartesian physics contained the world view of a plenum. In the 18th century this programme was concerned with hydrodynamics and with the mechanics of rigid bodies, and was fruitful in both fields (Euler, Bernoulli). In general, when considering two competitive research programmes it appears to be difficult to demonstrate that one of them is progressive, the other degenerative.

Therefore, it is perhaps interesting to consider successive rather than simultaneous programmes, for instance those of Copernicus and Newton, in order to investigate why Newton was successful, and Copernicus failed.

Copernicus' programme

In Chapter 3 we discussed the "Copernican program" of explaining motion by motion. At present we discuss "Copernicus' programme", which has a more limited scope. Copernicus wanted to explain the observed celestial motions by real uniform circular motion, with the sun motionless at the centre of the universe. Only later Copernicans extended this programme to all kinds of motions, replacing circular by linear inertial motion.

Copernicus' theory concerning the motion of celestial bodies may be considered a "research programme" in Lakatos' sense,[29] but then as an example that success is not always assured. His first model in the series was the final model all the same (see Secs. 2.4, 11.2).

The first model did nothing but suppose that the earth, like all other planets, travels around the sun, in uniform circular motion. This model enabled Copernicus to explain retrograde motion and some correlations with other phenomena. Moreover, he was able to calculate the relative dimensions of the planetary orbits (see Secs. 2.4, 2.5).

But that was all. Succeeding models employed epicycles and excenters, precisely in the same way as Copernicus' precursors had done. These tricks were used to describe the observed deviations from uniform circular motion, the first and third "inequalities" (see Sec. 2.1). Copernicus had to solve these problems by means of the methods of the competing, i.e., Ptolemy's research programme. Therefore, Lakatos calls this part of Copernicus' programme "degenerative."[30]

In the *Dialogue* (1632), the famous defense of Copernicanism, Galileo only discussed the first model of Copernicus' programme, keeping silent about the other steps. If Simplicio, who represents Aristotelianism in the *Dialogue*, had been free to choose his arguments, he would probably have pointed out that in the models following the first one, Copernicus shows himself an adherent of Ptolemy's methods, more than Galileo would like to admit (see Sec. 7.3).

Kepler also worked hard in the Copernican programme. Using Tycho's data, he tried but failed to go ahead even one more step. Ultimately he had to admit his failure, and his first and second laws showed Copernicus' programme to be mistaken. For this reason convinced Copernicans like Galileo ignored Kepler's achievements. Kepler was aware that his discoveries made a new research programme necessary, but he failed to start one. That was done by Newton.

Newton's work was generally considered the fulfillment of Copernicus' programme.[31] Discussing the solar system in the third book of *Principia*, Newton did not start from Kepler's first law, about the elliptic orbits. This law was a too radical departure from Copernicus' programme.[32] Instead, Newton started from Kepler's third law, the harmonic law, which is valid for circular as well as for elliptical orbits. Strictly speaking, Kepler's first and second laws refute Copernicus' programme. But the harmonic law is not only in full agreement with Copernicus' theory, it could only be found in the context of the Copernican theory, as we saw in Sec. 2.2. Therefore, Newton could afford to start with Kepler's harmonic law.[33]

But this was not sufficient. Newton had to make a completely new start, furthered by the newly developed mechanics. In other words, it is Newton's *synthesis* which, in a new research programme, finished Copernicus' programme.

Apparently with his tongue in his cheek, Newton observed that, if the "hypothesis acknowledged by all that the centre of the world is immovable" is true, the common centre of gravity of the solar system is immovable. But he hastened to add that the sun never recedes far from this centre.[34] For, as a true Copernican, he could not but adhere to the heliocentric view of the "system of the world", *i.e.*, the planetary system.

6.3. The method of analogy

The method of successive approximation is particularly suited for the investigation of the structure of matter. It is characterized by the fact that the successive models are increasingly complicated. Lakatos was so fascinated by his discovery of this method, that he did not see the existence of a reversed method.[35] In this method one abstracts as much as possible from concrete reality, in order to find universal laws, and universal modes of experience. It is applied whenever a scientist searches for unity.

Specification versus abstraction
The first of a series of models by which the study of the structure of matter proceeds is a rather crude and simple idealized approximation. It is intended to be gradually refined in the ensuing process. Hence, the method of successive approximation leads to an increasing specification. The structure to be studied is represented in ever more detail, but simultaneously the model becomes applicable to fewer systems. A rather crude model of the solar system may be applicable to the system of Jupiter with its moons, but the more we specify both systems, the more the models will become different.

If we are not first of all interested in the structure of matter but rather in general laws, we have to take an inverse route. Increasingly, we shall have to abstract from special circumstances, in order to get insight into universally valid relations. Because this route leads us away from concrete experience, the laws we discover will have an abstract character.

Analysis versus synthesis
In the study of the structure of matter, in which we want to approximate concrete reality step by step, we cannot restrict ourselves to a single aspect of that reality. For instance, studying the solar system, we cannot be satisfied to study merely numerical and spatial relations, mechanics and forces, but we also need optics,

magnetism, chemistry, geology, and so on. With an increasing complexity of the successive models we need ever more knowledge from various fields of science, in order to achieve the required synthesis.

The reverse study, looking for general laws, does not work synthetically, but analytically. In this study reality is pulled apart, analyzed, in order to find laws which are as general as possible. Its aim is not to gain full knowledge of the structure of a specific system, but to gain insight into the universal traits of all systems, irrespective of their typical structures.

Irreducible principles of explanation
Perhaps it would be nice if only one law existed to which the whole experienced reality would conform, but if such a law exists, it has not yet been found. It appears that a number of mutually irreducible aspects of experience or principles of explanation exist, each having a universal character. We mentioned and discussed four of them already in Chapter 3:

1. The *numerical* aspect, by which the Pythagoreans were so fascinated that they tried to explain everything in terms of rational numbers. Everything is related to anything else by numerical relations such as "larger than" or "as big as."

2. The *spatial* aspect, chosen by Plato to be the most important principle of explanation. Spatial relations like relative position have a universal character.

3. *Motion*, as an irreducible mode of experience put forward by 17th-century Copernicans. Recognizing "rest" as a state of motion implies that everything moves with respect to anything else.

4. *Interaction*, recognized by Newton to be irreducible to motion. The principle of relativity is a consequence of interaction being independent of time, position and external motion.

Without specification, these abstract principles do not define concrete structures of matter. Newton's second law, for instance, concerns an abstract and universal relation between force, mass and acceleration. If we want to apply this law to a system of bodies, we have first to specify the masses of the bodies, the forces acting between them, and a number of initial conditions, before we are eventually able to show that these bodies together form a system like Jupiter with its moons. As long as this specification is not given, Newton's second law cannot be applied. This is its weakness, but also its strength, its universal validity.

Besides these four aspects of human experience, we have already encountered the aspects of life, sensory experience, and logic. In the present chapter we discuss the historical or formative mode, and in chapters to come we shall deal with language, society, parsimony and harmony, justice, commitment, and belief.[36]

Analogy
There is a method for finding universal laws, but no more than the method of successive approximation does it guarantee success. No infallible method of finding laws exists. A universal law must be present in all concrete situations for which the law is supposed to be valid. The way to find such a law is to look for *analogies*.

Galileo used an analogy in his argument in favor of the earth's motion, or rather, in his explanation of the fact that this motion is not experienced. The relativity of motion is an abstract and universal principle, derived by analogy from the experience of people walking on a sailing ship. Galileo applied it to the motion of a cannon ball along the earth's surface.[37] He used a thought experiment concerning a ball falling from the mast of a sailing ship to explain why a ball falling from a tower does not deviate from the vertical.[38] Galileo explained the tidal motion by analogy with water transporting ships, "...those barges which are continually arriving from Fusina filled with water for the use of this city..."[39]

Gassendi was the first to consider gravity a force like any other, in this case as an attraction or something analogous to magnetic force.[40] Torricelli and Pascal derived their aerostatic laws from an analogy with hydrostatic laws. The abstract theory of waves is based on an analogy of water waves and sound waves, and is, according to Hooke and Huygens, applicable to light. The analogies between a falling body, a ball moving on an inclined plane, and a pendulum, played an important part in the development of 17th-century mechanics.

Newton's laws of motion can be defended only on analogical arguments. No experiment is possible to show that a body on which no unbalanced force is acting moves rectilinearly and uniformly; or to show that the acceleration of a body is inversely proportional to its mass and proportional to the acting force; or to show that action is negative reaction. The only way to find these laws and to make them plausible is to show that they are applicable in widely differing concrete situations, which structurally have little or nothing in common. Newton's laws of motion are applicable to planets, falling bodies, pendulums, mills, atoms, molecules, and electrons. The force concerned may be gravitational, magnetic, elastic, or electric — in any case, Newton's laws are valid, apart from relativistic corrections.

In the method of analogy, human fantasy plays perhaps a larger part than in the method of successive approximation. But it is an unwarranted thesis that the abstract universal laws are no more than the product of the human imagination, that is, no more than *conventions*. This thesis, defended by philosophers called "conventionalists", cannot be upheld if one realizes that these laws are applicable to concrete situations, open to experiment and observation. Although the results of these experiments never constitute a proof in a logical sense, it cannot be denied that even the most universal laws can be tested.

Frames of reference
The methodical abstraction described above does not only lead to universally valid law statements, but also to universal concepts and relations. This enables us to relate all kinds of things and events to each other, even if they have little in common. Concepts like force and mass, for instance, can be used in widely different situations.

In Newtonian mechanics, the concept of force is very prominent. "Force" is first an analogical concept enabling us to compare various types of interaction like gravity, magnetism and electricity in a *conceptual* sense. "Gravity is a kind of force, like magnetism." However, the analogy has also a physical meaning. Forces acting on the same body are able to balance each other, even if they are of a different kind. An electric force may balance the force of a spring. As a consequence we are able to compare forces of a different kind in a quantitative way — they can be measured. Because of the abstract, universal concept of force, we are able to compare widely differing phenomena, not merely in a logical, conceptual sense, but also with respect to their physical, kinematic, spatial and numerical relations.

The abstract modes of number, space, and motion provide us with a reference frame for events. The intensity of an event, the time when, and the position where, it takes place allow us to localize it, and to relate it to other events. During the Copernican revolution an important shift occurred in the views on space.[41] In ancient and medieval philosophy finite space constitutes a unique, concrete, differentiated continuum. The place of a body was determined by its immediate environment. Since the 17th century, infinite space is the abstract, undifferentiated three dimensional space of Euclidean geometry. Now the position of a body is determined with respect to an abstract coordinate system.[42] The principle of relativity shows that the choice of a coordinate system is arbitrary with respect to time, position, and uniform linear motion.

Hence, we see that the contrary directions of synthesis and analysis, or specification and abstraction, complement each other. In the direction of synthesis and specification, we approximate the concrete, the special, the complex, of which we want to find its typical *structure*. In the contrary direction we find the universal modes of human experience constituting a general and abstract framework connecting special and concrete things and events with each other in a *universal* way.

6.4. The method of mathematization

Another powerful heuristic, strongly furthered by Copernicanism, is mathematization, *i.e.*, the representation of kinematic or physical states of affairs in mathematical terms. Aristotelians did not believe the heuristic value of mathematics. Influenced by Pythagorean and Platonic ideas, both Kepler and Galileo have done much to promote the mathematical opening up of mechanics and physics.[43] Descartes, Huygens, and Leibniz were more mathematicians than physicists, and Newton's *Principia* was entitled the *Mathematical* principles of natural philosophy. His law of gravitation can only be rendered in mathematical terms.

It is striking that neither Popper nor Lakatos pays much attention to the historical fact that Newton found his theory of gravitation by means of a mathematical analysis. It suggests that mathematical analysis belongs to the "context of discovery" rather than to the "context of justification", a somewhat odd view. This does not imply that Newton's law is a mathematical law, or rests on purely mathematical arguments. The arguments are fully physical. Newton used mathematical methods in order to derive physical laws, dressed in mathematical formulae, from other physical laws.

Principia Mathematica
Newton's *Principia* (1687) is the most prominent example of the fruitfulness of mathematics in physics. Besides the introduction to mechanics (see Sec. 3.6), the *Principia* consists of three books. The first and second books are mostly mathematical treatises, the first concerned with motion in a vacuum, the second with motion in a material medium. The third book, preceded by Newton's *Rules of Reasoning in Philosophy*, contains his "System of the World", and

is truly physical or astronomical. It is concluded by a General Scholium.

In his youth, Newton had carefully studied Descartes' works, in particular his *Principles of Philosophy* (1644). In this book Descartes propagated the study of nature by the method of mathematics, but Newton noted that Descartes did not keep his promise. Descartes' *more geometrico*, the geometrical way, means that physics should proceed axiomatically, like Euclid's geometry, starting from clear and distinct ideas. Descartes' physics was not mathematical according to Newton's standards. In the second book of the *Principia*, Newton criticized Descartes' theory of vortices by showing mathematically that it contradicted Kepler's laws of planetary motion (see Sec. 10.2).

In the first book, Newton carefully distinguished the mathematical principle of "force" from its physical meaning. Mathematically he derived how large a force must be, and which direction it must have, if under its influence a body is to move in an elliptical path. But it is a *physical* matter to decide in any particular case whether this "centripetal force" is gravitational, elastic, electric, or magnetic, and what the nature of these forces is. The physical aspect of the gravitational force was only considered in *Principia*'s third book.

The inverse-square law

It is not exactly known how Newton found the law of gravity, and the following is a "rational reconstruction" (see Sec. 11.2) which, however, is based on Newton's own account in the three consecutive editions of *Principia* (1687, 1713, 1726).[44] In the third edition Newton expounded his heuristic in the so-called *Regulae Philosophandi* — Rules of reasoning in philosophy.[45]

The first rule is: "We are to admit no more causes of natural things than such as are both true and sufficient to explain their appearances." Therefore, Newton assumes that only one force operative in the solar system is sufficient to explain the curved orbits and changing velocities of the planets. Newton also assumes that the same law must be valid for the solar system and for the planets having satellites, according to the second rule: "Therefore to the same natural effect we must, as far as possible, assign the same causes."

From Kepler's second law, the area law, which Newton derived from his own third law, it can be found that the force must be centripetal, *i.e.*, directed to a fixed point.[46] Several orbital shapes are consistent with this law, among them circular orbits. In that case, Kepler's second law says that the orbital speed is constant. For uniform circular motion Huygens had derived the magnitude of the centripetal acceleration as a function of the radius R and the speed v.

Now Newton considers a number of mass points moving in hypothetical circular homocentric orbits with different radii R and periods T. He applies Kepler's third law (R^3/T^2 is constant) and Huygens' formula for centripetal acceleration ($a = v^2/R = 4\pi^2R/T^2$) in order to show that the acceleration is inversely proportional to R^2. This straightforward part of Newton's derivation was also found by Wren, Hooke, and Halley, and probably several other scientists.[47]

Mass
According to Newton's second law ($F = ma$), the force by which each hypothetical mass point is drawn to the centre is therefore proportional to its mass m and inversely proportional to the square of the distance R.

Next, Newton applies the third rule of reasoning: "The qualities of bodies, which admit neither intensification nor remission of degrees, and which are found to belong to all bodies within the reach of our experiments, are to be esteemed the universal qualities of all bodies whatsoever", on which rule Newton comments: "...we must, in consequence of this rule universally allow that all bodies whatsoever are endowed with a principle of mutual gravitation." Combined with the third law of motion, this means that if the hypothetical mass point (say, a planet) is attracted to the centre (say, the sun) by a force proportional to the planet's mass (see above), then the sun is attracted by the planet with an equal force, now proportional to the sun's mass. Hence, the force between the sun and the planet is proportional to the mass of the sun *and* the mass of the planet, and is inversely proportional to the square of their mutual distance.

This second part of the derivation of the law of gravity is completely due to Newton.

Generalization
Finally, Newton generalized this law, found for the ideal case of uniform circular motions, to all kinds of motion influenced by gravitational interaction, elliptical non-uniform orbits, projectile motion, free fall, and pendulum motion. This is in accord with the fourth rule, reading: "In experimental philosophy we are to look upon propositions inferred by general induction from phenomena as

accurately or very nearly true, notwithstanding any contrary hypotheses that may be imagined till such time as other phenomena occur, by which they will either be made more accurate, or liable to exceptions."

This generalization is an act of the imagination. Newton assumes that the force responsible for the motion of the planets is the universal force of gravity. He calls this act "induction", and rightly, though it is performed at a higher level than that of "empirical generalization." It is based on the belief that natural laws have universal validity. Therefore, if it is valid in one particular case, it should be valid in other cases as well.

After the derivation of the law of gravity Newton turns to the problem of the solar system, which he partly solves by the method of successive approximation described in Sec. 6.2. According to deductivists, including Lakatos, Newton's scientific work started here. The abstracting phase, unmistakably present in the above story, and the mathematical derivation of the law of gravity, are considered not to be a matter of logic, but to belong to the "context of discovery", the domain of history or psychology.

Granting that this story represents an important piece of history, showing how theories are formed, we still maintain that this phase is not much less logical or rational than the succeeding deductive phase, which is considered a paradigm of logical reasoning. The way by which Newton found his law of gravitation shows no trace of the method of successive approximation which Lakatos elevated to the standard method of scientific research.

Deductivist criticism and feed-back

Deductivists state that Newton's derivation as related above cannot be logical, because it starts from Kepler's laws, which are subsequently demonstrated to be false (see Sec. 5.1). This argument can now hardly be taken seriously if the same philosophers praise Newton for the remainder of his work. For Lakatos' method of successive approximation also can only be fruitful on the basis of idealizing statements concerning the structure of the solar system, which are shown to be false in the next approximation.

We should rather consider Newton's method to be an instance of control by feed-back. If we consider Kepler's laws as the initial input of the feed-back system, and Newton's law of gravity as its output, we find that the output corrects the input. This is very familiar in feed-back systems, and often leads to very stable results.

Inertial and gravitational mass

We wish to comment on the widely accepted fable that Newton made a distinction between inertial mass, the concept occurring in the second law of motion ($F = ma$), and gravitational mass, the concept occurring in the law of gravity ($F = Gm_1m_2/R^2$)

Newton never made this distinction. He did distinguish between mass, defined as quantity of matter or *vis inertiae*, and weight, identified with the force of gravity.[48] But weight, as a force, can never be identified with gravitational mass, occurring in the law of gravitation.[49]

In fact, Newton could never have made the distinction between "inertial" and "gravitational" mass, because he used the second law of motion, in which the "inertial" mass occurs, in order to show that the force exerted by the sun on a planet is proportional to the planet's mass. Hence, the mass occurring in the law of gravity *is* the "inertial" mass.

The fable that Newton did make this distinction has come to life probably after the discovery of similar inverse-square laws for electric and magnetic forces. For instance, in Coulomb's law of electric force, electric charge plays a similar role as mass in Newton's law of gravitation. Only then people started to consider mass the source of gravity, like electric charge is the source of electricity.[50] But this view is contrary to Newton's, because it overlooks the way Newton obtained his law.

The view that "inertial" and "gravitational" mass must be distinguished fits Popper's hypothetical-deductive method of conjectures and refutations very well.[51] According to this view the second law of motion and the law of gravity are independent conjectures, found by trial and error. Then it is an empirical fact that "inertial" mass turns out to be proportional to "gravitational" mass, such that they may be equated for practical reasons. But this view is historically untenable, and logically very unlikely.

6.5. The function of technology

Mathematics may be considered a science founding the natural sciences, because mathematics is applied in, *e.g.*, physics. Not only the natural sciences benefit from this relation. In the 16th and 17th centuries mathematics was developed remarkably, often stimulated by problems taken from mechanics or crafts like bookkeeping.[52] Decimal fractions came into use at the end of the 16th century, when also logarithmic calculation was invented. The

17th and 18th centuries have seen many great mathematicians who also did important work in physics: Fermat, Descartes, Pascal, Huygens, Newton, Leibniz, Euler, the Bernoullis, *etc.* The development of differential calculus was summoned by problems in mechanics.

In its turn physics is foundational to chemistry and biology, and these sciences frequently make use of physical results. But the most important field of application of physics is technology. Roman and medieval culture witnessed important progress in technology. Until the Renaissance, the arts were developed completely independent of academic science, however (see Sec. 8.2). Only in the 16th and 17th centuries were technology and science tied together. Initially technology remained practically independent of science, whereas physics and chemistry increasingly relied on instrumental techniques.[53] Only after about 1850 did technology begin to benefit from scientific research. Up till then science owed more to technology than *vice versa.*

At present we shall mainly be concerned with the heuristic value of instruments for the development of science.

Tycho Brahe's astronomical observations

The first requirement for the application of mathematics to physics is the availability of physical *magnitudes*, which values can be determined by measurement. The significance of accurate measurements was clearly recognized by the Copernicans. Already in the 16th century, the existence of two rivalling theories, Ptolemy's and Copernicus', showed the need for improved observational methods.

Copernicus himself did very few measurements. In particular the great Danish astronomer Tycho Brahe took up the challenge.[54] Supported by his own resources and those of the Danish king, he built *c.*1580 a modern observatory, which became a model for later observatories (see Sec. 8.1). Tycho was not only an able astronomer, but also a competent though despotic organizer. The observational instruments which he designed were the best ever made before the invention of the telescope. They were not only very precise, but were even supplied with a table of errors.

Tycho's innovations now look simple, but in the 16th century they were revolutionary. First, he considerably increased the accuracy of the measurement of stellar positions. Until Copernicus people were content with an inaccuracy of 10', about one third of the moon's apparent diameter. Tycho reduced this value to 2', sometimes to 1' or 0.5'. Tycho's precision was not improved until seventy years after his death.

Secondly, Tycho's observations were far more systematic than before. He determined systematically the positions of several hundred visible stars. Moreover, he measured the positions of the planets every clear night. Earlier measurements of planetary motion were often confined to the stationary points of a retrogradation and similar striking events, like the conjunction of two or more planets. Tycho's results were published by Kepler in the *Rudolphine Tables* (1627).

Astronomia Nova

Without Tycho's measurements, Kepler would never have found his laws, published in *Astronomia Nova* (1609). For a long time, Kepler tried to describe the motion of Mars according to Copernicus' programme, *i.e.*, with the help of uniform, circular motions, and he nearly succeeded.

In accord with contemporary methods, he tried to obtain correspondence between the theory and the measurements with respect to the positions and velocities of the planet at four points. These points were the perihelion and aphelion (the points of minimum and maximum distance to the sun) and two points in between. After arriving at a quite satisfactory result, Kepler did something nobody had done before. He checked whether the solution obtained was also satisfying at other positions, and he found a discrepancy between observed and calculated positions of 8'. This value corresponds to about a quarter of the moon's apparent diameter.[55]

Before *c.*1580 the precision of astronomical measurements was seldom better than 10'. Hence, Kepler's results would have seemed quite satisfactory to Copernicus and his contemporaries. But Tycho's measurements, the basis of Kepler's calculations, achieved a precision of 1' or 2', which meant that Kepler's theory was proved wrong. He wrote: "But for us, who, by divine kindness were given an accurate observer such as Tycho Brahe, for us it is fitting that we should acknowledge this divine gift and put it to use...Henceforth, I shall lead the way towards that goal according to my own ideas. For, if I had believed that we could ignore these eight minutes, I should have patched up my hypothesis accordingly. But since it was not permissible to ignore them, those eight minutes point the road to a complete reformation of astronomy: they have become the building material for a large part of this work....."[56] This work included his first two laws, the elliptic law and the area law.

The telescope

The 17th century witnessed the increasing ability of craftsmen like shipbuilders and instrument makers. Many inventions date from the 17th century, such as the pendulum clock, the thermometer, the barometer, and the microscope. For 17th-century astronomy the invention and development of the telescope was of foremost importance. About 1600 several instrument makers in Italy and the Netherlands experimented with combined lenses. In 1609 Galileo heard a rumour about the results of a Dutch artisan, and being an instrument maker himself, he succeeded in making telescopes which for at least a decade were the best available.

Galileo was the first to use the telescope in order to investigate the heavens. In 1609-10 he made his great discoveries — mountains on the moon, the sunspots, Jupiter's moons, Saturn's "ears", the phases of Venus, and many new stars. After Galileo, the next interesting discoveries were due to Huygens, who explained Saturn's "ears" to constitute a ring, and who discovered the first moon of the same planet.

Because of his discoveries, Huygens claimed to have the best telescope of Europe. Whether true or not, he was soon surpassed by G. D. Cassini at Paris who observed the sky with telescopes made by the Italian instrument maker G. Campani. Until 1685, the most important discoveries, including some moons of Saturn, were made by Cassini. Huygens' contribution to telescope design consists of the inventions of the micrometer, a new ocular, and the application of cross-hairs. Also the invention of the pendulum clock improved the accuracy of astronomical measurements, even improving the standard set by Tycho.

In England, in particular by Flamsteed, the first director of the Greenwich Royal Observatory, important contributions were made to the technology of telescopic observation.

The heuristic value of accurate measurements

The heuristic value of accurate measurements can hardly be overestimated. Kepler would never have found his laws if he did not have available Tycho's observation results, known to be much more accurate than any observation before. Newton used the measurements of Kepler, Cassini, Borelli, Flamsteed and other astronomers as arguments in favour of his theory of gravitation. Galileo made accurate measurements on balls moving down an inclined plane, in order to arrive at arguments for the derivation of the law of fall.[57] Newton carefully measured the periods of various pendulums in order to establish that gravity is independent of the density of the oscillating bodies, and to determine the velocity-dependence of friction.[58]

From a careful measurement of the periods of Jupiter's moons at different dates, Rømer (1676) concluded that the speed of light is not infinite, as Descartes had stated (see Sec. 11.3). From his result, Huygens calculated the value of the speed of light for the first time in history.

Torricelli's and Pascal's precise measurements with barometers laid the foundation of modern views on the pressure of air. Pascal's prediction that the pressure would be decreasing at increasing height, and the experiment carried out by his brother-in-law at the Puy-de-Dôme (1648) made a deep impression among the European learned.

In 1672, Richer demonstrated that the weight of a body depends on its position on earth, in particular on the latitude. He observed that a clock calibrated at Paris was 2'28" per day slow at Cayenne, and he argued convincingly that this could not be ascribed to the thermal expansion of the pendulum. Newton used this result in order to show the relevance of making a distinction between mass and weight.[59]

These examples may suffice to show that accurate measurements have not merely a function in the testing of theories, in the "context of justification." They may have an important heuristic function, in the "context of discovery", and this function has become ever more important since the Copernican revolution, when it was born.

6.6. Theory of the opening process

In Secs. 4.4 and 4.5 we observed that during the Copernican revolution the aim of science received a new emphasis. In particular during the middle ages, scientific activity was mostly a logical affair, consisting of the logical analysis of the theories of Aristotle and other authorities. The logical-empiricists were probably not aware that they affirmed this medieval view by distinguishing between a "context of discovery" and a "context of justification", recognizing only the latter as a rational affair. In the present chapter we have tried to show that the historical development of science is by no means irrational. Rather, we state that since the Copernican revolution science is more historical than logical, because the aim of science is to develop the lawful structure of nature.

By way of summary, we shall now discuss an explanatory

model of the "scientific opening process."

Pluralism of methods

We have seen that there is more to be said about scientific method than a discussion pro or contra induction. It appears that the views of the logical-empiricists on induction and the relation of logic, theory and observation are rather poor, just like Popper's method of trial and error, or conjectures and refutations. This has been pointed out by Feyerabend. He observes that scientists hardly ever work according to the views of logical-empiricists, of Popper, or of Lakatos. On the contrary, scientists use any means to achieve their goal. Feyerabend speaks of a pluralism of methods, even about an anarchy.[60]

Unfortunately, it is not quite clear what he considers to be the aim of science. We consider this aim to be the opening up of the law side of nature, the discovery and development of law conformity in reality. Indeed, the scientist has available a great diversity of methods, but these do not display an anarchy. Analysis and synthesis are complementary, just like the mathematical and technical or instrumental opening up of a field of science. We discussed these four methods separately, but in our reconstruction of Newton's derivation of the law of gravity we easily recognized more than one method at work.

An able scientist masters all methods of his discipline, and to tackle a problem he chooses the methods which suit him best.[61]

Linearity or complementarity

The views of Popper, Lakatos and others about the progress of science are frustrated by their implicit supposition that the development of science is a *linear* process. This view is an inheritance of logical-empiricism, which considered scientific progress to be a continuous increase of knowledge, an accumulation of theoretical insights or of understanding, as well as of simple facts and known laws. It should be criticized on two accounts.

First, the development of science must not be conceived as a linear process, but as a process in several dimensions, in which every direction has its own heuristic.[62] We argued that the methods of successive approximation and of abstraction are complementary, but proceed in opposite directions. Also the mathematical and instrumental heuristics may be considered opposite, because in principle (though not always actually), instrumentation depends on physics, and physics depends on mathematics. Finally, Popper notwithstanding, deduction and induction may very well be considered to be opposite means of developing our knowledge of laws.

Secondly, we observe that Lakatos' competitive research programmes are often if not always determined by incompatible *world views*, in which the hypostatization of a certain aspect of reality belongs to the "hard core." We point to Descartes' hypostatization of the spatial and kinematic aspects, or Newton's force-matter dualism. The one-sided emphasis on certain principles of explanation may lead to stagnation, but also to an increased effort to deepen the principle concerned. This means that rival research programmes may alternatively be "progressive" and "degenerative" in Lakatos' sense, because some problems can better be solved starting from one principle than from another, or with the help of one heuristic rather than with another.

However, this implies that a choice in favour of one research programme and the consequent rejection of its rival is not objectively possible. Either one makes a choice between various world views, or one recognizes that the hypostatization of a single method or principle of explanation is to be rejected. In the latter case a scientist is free to use all available principles of explanation, and any method he thinks fit to solving his problems, being aware of the fact that no single method is sufficient to solve all problems.

Three basic distinctions

Our explanatory model of the scientific opening process is based on three distinctions, introduced by the Christian philosopher Herman Dooyeweerd.[63] The first distinction is that between laws and anything that is subject to laws. All things, events and relations are subject to laws, and it is the aim of science to discover and investigate the lawful structure of the world. This distinction gives rise to philosophical questions like that about the relation between laws and their subjects, about the status of theories, statements and concepts, their meaning and corroboration, about deduction and induction. It also leads to the question of the origin of laws, and the answer to this question, if explicitly given, betrays one's faith. In a (neo-) Kantian philosophy, for instance, the answer will be, that laws are given by man, in order to make sense of his sensations. In a Christian philosophy, the answer will probably be that the laws are given by God, who uses them to constitute, govern, and maintain his creation. The laws are not merely *descriptive* (describing regularities), but also *prescriptive, i.e.,* given by a Lawgiver. Hence, in a Christian view, laws are not given or proved by man (contrary to law statements), but have to be discovered in a careful and respectful exploration of the creation (see Chapter 12).

The second basic distinction is made between universal, general modes of being or experience, and typical or structural ones. Typical laws determine the structure and functioning of things and events as far as they differ according to specific properties. Universal laws, on the contrary, concern general relations between things and events. Whatever the typical differences between various kinds of entities, individual entities are always related by their spatial distance, relative motion, and mutual interaction.

The third basic distinction concerns the mutual irreducibility and relatedness of the various modes of human experience. For the physical sciences, four of these are most important: the numerical, the spatial, the kinematic and the physical ones. In this order, they are followed by the biological, the sensory, the logical, and other modes of explanation. These modes are related to each other, first because they show a linear order, next because each mode refers to the other modes. In this linear order, we can speak of referring forward (anticipations) or backward (retrocipations).

We consider these basic distinctions to be mutually independent, like the axes of a three dimensional coordinate system.[64] If we relate the first distinction to the vertical axis, the "upward" z-direction points to the lawside of nature, the "downward" z-direction to the subject side. The second distinction (say, the y-axis) is also a dichotomy, distinguishing the universal, "right" side from the concrete, typical or individual "left" side. The third axis, the x-axis, displays the series of irreducible modes of experience mentioned above. This picture suggests that the basic distinctions define three pairs of "directions" of research. It shows that research is not a linear process, but a multidimensional unfolding one.

Whereas the first basic distinction is of special interest to *philosophers* of science, the other two determine the opening process, and hence are very important for the understanding of the *history* of science, or rather of science as history. In the scientific opening process we identified four "directions of research", besides the processes of "induction" (directed to the law side) and "deduction" (directed to the subject side).

Two directions are characterized by the second basic distinction, the distinction between universal and typical laws. They concern the search for *unity* by the method of analogy, and the search for *structure* by the method of successive approximation. The other two directions are the search for *objectivity* by means of mathematization, and the search for *application*, by means of instrumentation and the development of artifacts like machines. These are determined respectively by the retrocipations (backward relations) and anticipations (forward relations) in the linear order of the various modes of experience.

We shall briefly summarize these four directions of research.

The search for objectivity

Both Descartes and Newton saw the fundamental aim of science to be the provision of an objective description of natural states of affairs in mathematical terms. In our view, objectivity means the same. Objectivity in the natural sciences means the representation of physical states of affairs in mathematical formulas. This includes the view that objectivity in physics means the formulation of physical laws independent of time, space and motion, because the possibility for this principle of relativity follows from the mutual irreducibility of the physical and the mathematical modes of experience. In our explanatory model, the numerical, spatial and kinematic aspects of reality precede the physical one. Hence, objectivity is related to the above defined retrocipatory direction in the opening process.

For mathematization it is necessary to have quantitative concepts concerning directly or indirectly measurable properties of things and events. The metric (the scale with unit) of such a property must be defined, and the meaning of all concepts in a mathematical theory of a field of science must be determined and mutually related by a number of laws.

The search for application

Besides the search for objectivity, an important motive for many Copernicans to develop a mathematical theory of mechanics was to explore its practical use. This is traceable to Bacon's influential writings on the philosophy of science. Often, the same people who worked on the mathematical theories also invented instruments — for instance, Galileo, Descartes, Huygens, and Newton. They were not only concerned as scientists with nature, but also concerned as technicians with the construction of artifacts. The explanation of an artifact cannot be restricted to its mathematical and physical aspects, but must be extended to its *use*. Artifacts are always constructed with some purpose, which is not found in nature.

In particular, measuring apparatus is immediately related to the purpose of the mathematization of science. Telescopes and pendulum clocks are artificial things, made with a certain aim — to improve astronomical observations, for instance — and thereby are useful for the development of theories. Besides, they have important practical use, outside science.

Any physical instrument anticipates the *biological* modal aspect, because it is an organized whole of deliberately connected parts, each having a certain function. Besides, any instrument must have some observational property, anticipating the *sensory* mode of experience. Its function must have some *logical* relation with a theory about the meaning and the metric of the property to be measured. Finally, it has been invented and designed by *historical* persons.

The dynamic motif of technological research is, of course, utility. Hence we can grant pragmatic philosophers (including Marxists) that "praxis" is a very strong motif in scientific research.

The search for universality

The search for unifying, universally valid and general laws is especially inspired by people who seek the unity of nature within nature itself. This search has been highly successful in the development of mechanics, which investigated the kinematic aspect of reality considered irreducible to the numerical and spatial modes of experience (see Secs. 3.3, 3.4).

It has also led to attempts to reduce all physical experience to a single mode, in particular by the mechanists, who tried to reduce all physical phenomena to matter in motion. We have seen that Newton's introduction of "occult" ideas like action at a distance was met with great suspicion by the Cartesian mechanists (see Secs. 3.5, 3.6).

The search for structure

The introduction of generally valid abstract concepts and laws is not the only means to find unity in the physical sciences. The reverse method is to apply a number of fields of science to the same typical problem, for example, the solar system. In this synthetical direction, experimental physicists play a larger role than in the other three. Kepler would not have found his laws if he did not have Tycho's careful measurements at his disposal. More often than not, one has first to purify a sample of the structure to be studied, and then to collect a number of data, before a theory can be suggested. Such an initial theory or model will then be helpful to make purer samples, and to design new experiments, leading to a more sophisticated theory. Hence, the interplay of theory and experiment comes more to the fore in this kind of research than in the former three. This direction of research is characterized by the method of successive approximation, as we have seen, but also by the synthesis of various fields of science (see Sec. 5.1).

The natural order

This explanatory model with its three basic distinctions tries to account for the empirically discovered natural order. (It is a matter of belief, of course, that this order is created.) In the past few decades, several philosophers and historians of science have argued that science develops itself into a certain direction, guided by a paradigm (Kuhn), a research programme (Lakatos), or a research tradition (Laudan). They have shown the intimate relation between a metaphysical world view and scientific work. But none of them has recognized the *natural order* of these various directions of scientific research.

Assuming that any research programme, if it is successful, must be bound to this natural order, our explanatory model identifies four different directions of research. The model can explain why certain world views are more successfully working in one direction rather than in another. It can explain why the sometimes vehement discussions between adherents of different world views never have succeeded in repressing one of them. It explains why different points of view are necessary to obtain a fully developed insight into nature.

7. The principle of clarity

7.1. Communication

To this point we have studied the structure and functioning of a theory, its relations to other theories, and the way theories are discovered. We emphasized that theories are used by people with differing aims, different views about the functions of theories, and with various views on their reliability.

In the chapters to come we shall be concerned with some normative aspects of theorizing: how theories are made clear, how they function in society, their relation to the principles of parsimony and harmony, how they are judged, to which degree one should be committed to one's theories, and finally, the question of truth.

In the present chapter we turn to dialectic or rhetoric, the art of the argumentative usage of language,[1] *i.e.*, the way people converse with each other when they use theories and other logically qualified structural wholes. Rhetoric is subject to the norm of *clarity*. It concerns the question of how theories are communicated, how they are made clear to various people.

Is science a language?
The question whether science is a language is answered in the affirmative by many people.[2] They consider concepts to be symbols, such that there is little difference between a word and a concept. The aim of science is "hermeneutic", to translate reality, to interpret it, to clarify it.

Also the logical-empiricists, in particular the so-called language-analysts, are attached to this view. Assuming that ordinary languages like Latin, German or English are not suited for this purpose, they proposed that science should develop its own formal language, unequivocal, and interpretable in only one way. Because they distinguished among logic (including mathematics), theory, and observation they assumed the need of three vocabularies, a logical, a theoretical, and an empirical one.[3] An important part of philosophical discussions concerns the possibility of reducing theoretical language to observation language, *i.e.*, of translating all theoretical statements into observation statements, supposed to be independent of any theory[4] (see Sec. 5.3).

We agree that language has an important hermeneutic function in science, but maintain that science itself is not linguistically qualified.[5] We have already seen that science has instead a historical function — the progressive opening up of the lawful character of reality (see Secs. 4.4, 4.5, 6.6). In the present chapter we restrict ourselves mostly to the investigation of the role of language in 17th-century science.

The main function of language is communication, the transfer of feelings, thought and skills by means of symbols and metaphors.

Levels of communication

Apart from education, to be treated in Sec. 7.2, scientific communication[6] occurs at three levels, first, between specialists on the same subject, next, between specialist and non-specialist scientists, and thirdly, between scientists and non-scientists. All three have specific requirements with respect to the use of language. Especially on the first mentioned "highest" level we often meet a specialized technical terminology. The middle level is very important to make the results of one specialization available to others. If this intermediate level does not function satisfactorily, the network character of theories is endangered, and stagnation may follow.

In the 17th century, the basis was laid for this threefold communication system. At the highest level, communication took place in Latin, by means of books and letters. The distribution of books was much better organized than during the middle ages, thanks to the invention of printing in the 15th century. But the number of copies of each printed book was relatively small, not more than a few hundreds every edition. Copyright did not exist, and often books were reprinted without permission of the author or the first publisher.

Both Copernicus' *Revolutionibus* and Newton's *Principia* were written for specialists. In his Preface, Copernicus emphasized that "mathematics is written for mathematicians."[7] At the beginning of *Principia's* third book, Newton says that he had abandoned the plan to write it in a popular way, "...to prevent the disputes which might be raised upon such accounts,"[8] *i.e.*, to avoid the criticism of incompetent scholars (see Sec. 10.3). Nevertheless, he indicates which parts of the book can be omitted at first reading. During Newton's lifetime, the *Principia* was not translated into English, though it was into French. His *Opticks* (1704) is a far less technical work, and was initially written in English (later also in Latin).

Another channel of communication at the highest level was the exchange of letters — by Kepler and Galileo, by Descartes and Mersenne, by Newton and Bentley, by Clarke and Leibniz. Often, these letters were copied several times, and many of them have been preserved to form an important source of knowledge about 17th-century science.

At the intermediate level, an important novelty was the introduction of scientific journals, for instance, the *Philosophical Transactions* of the Royal Society at London. Nowadays these journals have a specialist character and hence belong to the first category. But in the 17th century their aim was to inform nonspecialists of the proceedings of science. Initially most papers were written in Latin, but gradually Latin had to give way to the common languages: English, French, German, Italian, and Dutch.

At the third level especially Galileo did pioneering work, by writing his most influential work in colloquial Italian, in such a style that it was readable by non-scientists, that is, for an intelligent public. This contributed to Galileo's popularity as well as to the dispersion of new views. In the Netherlands, Galileo was preceded by Simon Stevin, who pleaded for writing scientific works in common language in a comprehensible way. According to Stevin, "...the Greeks were of the most intelligent that Nature produces, but they lacked a good tool, that is, the Dutch language, without which in the most profound matters one can accomplish as little as a skilled carpenter without good tempered tools can carry on his trade." (Quoted by Drake,[9] who adds: "Galileo was later to say that the book of Nature was written in mathematics. It seems that for Stevin the book of Nature was written in Dutch".) An example is his *Weeghconst* (The art of weighing, 1586), a book about mechanics.

After Galileo, Descartes wrote his influential *Discourse on Method* (1637) in French. "And if I write in French, which is the language of my country, rather than in Latin, which is that of my teachers, it is because I hope that those who rely purely on their natural intelligence will be better judges of my views than those who believe only what they find in the writings of antiquity."[10] The *Discourse* is a popular exposition of Descartes' philosophy.

His scientific work was *Meditationes de Prima Philosophia* (1641) and *Principia Philosophiae* (1644), both written in Latin, though the latter was shortly afterwards translated into French.[11]

Except for Galileo,[12] no leading scientist of the 17th century wrote really popular works, leaving that to their disciples, usually scholars of high quality but lesser creativity. Thus Rohault popularized Descartes, and Clarke and Voltaire did the same for Newton.

The popularization of scientific results strongly influences the common world view. The Western world view has changed radically since the Copernican revolution, though it took a long time before the idea of a moving earth became commonly accepted. In the 17th century even the majority of the learned remained convinced of the geocentric world view, and discussions about the validity of the heliocentric system continued deep into the 18th century.

Mathematics as language of physics
It is an often heard statement that mathematics is the language of physics. This superficial view is clearly contradicted by the view that physics itself is a language. If one rejects this, and mathematics is accepted as a science on a par with physics, one should also reject the idea that mathematics is a language. Only if one considers mathematics to be fundamentally different from science could one accept the statement that it is a vehicle of science.

In *Il Saggiatore* (1623) Galileo seems to adhere to this view:[13] "Philosophy is written in this grand book, the universe, which stands continually open to our gaze. But the book cannot be understood unless one first learns to comprehend the language and read the letters in which it is composed. It is written in the language of mathematics, and its characters are triangles, circles and other geometric figures without which it is humanly impossible to understand a single word of it; without these, one wanders about in a dark labyrinth."

In this passage, Galileo reveals his neo-Platonic background according to which nature can only be explained with the help of "mathematical" ideas. Galileo's mentioning a "language" is merely a metaphor.

In Sec. 6.4 we proposed a far more intricate relation between physics and mathematics. Because the kinematic and physical aspects of experience are founded by the numerical and spatial ones, quantitative and spatial aspects play an important part in the search for objectivity in physics and mechanics. Therefore, mathematics as the science of quantitative and spatial relationships constitutes one of the most important presuppositions of physical theories. This includes physics making use of the language of mathematics.

The language of mathematics itself has been developed during the 17th century in a way that has been of the greatest importance also for physics. This concerns the introduction of the algebraic symbol system, the use of letters as symbols for numbers and variables. This is indeed an example of the introduction of a typical language into science, showing how important and fruitful a right and well-considered use of language can be. Another example is the introduction of decimal fractions by Simon Stevin, although a hundred years later Newton still applied ordinary fractions. Finally, formulas as expressions of physical laws and relations came into use only very slowly. Started in the 17th century, it was commonly applied only in the 19th century.

7.2. Didactics

In the preceding chapters we have seen that the principal function of a theory is to allow people to discuss the truth of statements. Each theory has a logical function in a logical debate between several disputants. The participants in the debate have to agree about the starting points (otherwise the discussion is impossible), and to disagree about the theorems to be discussed (or else the debate is superfluous).

The function of language is to achieve clarity about all this — the starting points, the debated issues, the data, the argument, the results, the tests.

Language and logic
Language should not be confused with argument. An argument can be very clear, yet not convincing. Obversely, reasoning may be accepted even if not all its parts are completely clear. Often it takes a long time before a theory acquires its maximum clarity. This was, for instance, the case with Newton's mechanics, which only in the 19th century was to reach an appreciable level of clarity. The unequivocal meaning of the concept of force, in particular, was not achieved before the middle of that century.

The strength of an argument is enhanced if it is easily understood and readily accepted. Contrary to his first and second law, Kepler's third law was both easy to understand and uncontroversial. This may have been another reason why Newton accorded it a prominent position in his argument concerning the solar system (see Sec. 6.2). Galileo's argument in favour of Copernicanism, derived from the apparent motion of the sunspots, is even today difficult to understand, and convinced nobody.[14]

Argumentation needs a special kind of language, in two respects. First, it requires its own logical indicators, words like "therefore, thus, so, hence, consequently, because, since, for."[15] Secondly, any science is in need of its own language. The logical requirement of having unequivocal concepts is at variance with the lingual variety of homonyms and synonyms (which, on the other hand, serves the needs of didactics very well). For instance, while acceptable in an Aristotelian context, the homonymic use of the word "earth", meaning "soil", "the heaviest element", or "the globe", became for a Copernican like Galileo an unbearable equivocation.[16] Copernicanism had to introduce new words, or rather new meaning to existing words, like force, mass, quantity of motion. In this process, science finds itself withdrawing from common language.

Language is largely characterized by the use of metaphors. In science, metaphors are used to clarify theories. The choice of metaphors is strongly determined by one's world view. Thus, during and before the middle ages, the organistic world view required mechanical instruments like the lever to be made clear by comparing them to living systems. In a mechanist world view, it is the other way round. Now the functioning of the human arm is explained by comparing it with a lever. Since the Copernican revolution, the clockwork metaphor became very popular in clarifying the new system of the world.[17]

Education

Theories must be explained because they deviate from common sense. "To explain" has a logical and a lingual meaning. Its logical sense was discussed in Chapters 2 and 3. In a lingual sense, "to explain" means to clarify, to take away any barrier to understanding.

It is a didactic aim to clarify theories, to make people understand them. It turns out that to different people the same theory must be clarified in different ways. The amount of clarification needed depends on their background knowledge, their experience, their intelligence, their mastering of language, and their willingness to understand what is told.

Because a theory is an instrument, for instance to solve problems, the use of a theory must be exercised. In exercises the mastering of the language used is enlarged. On the one hand, a student learns to apply his own language to new problems, and on the other hand he extends his language, by learning and applying new words and expressions. In this way a student not only gains clarity about a theory and its possible applications, but also about that part of reality with which the theory is concerned.

For the large part, medieval "scholastic" education consisted of the citation of authorities. In the discussion of a thesis, a long list of arguments, taken from the literature, was presented, both pro and contra the thesis, and in the end one's own view was given. The best student was the one able to quote most citations, preferably from memory. Whereas Kepler's *Mysterium Cosmographicum* (1597) and *Astronomia Nova* (1609) are scientific works, his *Epitome Astronomiae Copernicanae* (1617-21) is a summary and a textbook. Consequently, Kepler adopted the usual form of question and answer, gave a systematic pedagogical order to his expositions, and avoided biographical details.[18]

The turn of the tide came during the Copernican revolution. The French scholar Ramus advocated the educational value of visiting artisans, which he valued more than scholastic training. This means that he favoured the study of skills more than the study of bookish knowledge.

Galileo also attacked the scholastic way of education.[19] He preferred to give his students new problems which they should try to solve, if possible with new arguments derived from common sense, observation, and experiment. In his *Dialogue*, Galileo's spokesman Salviati challenges his opponent Simplicio to forget about Aristotle, and to use his own wits. This means that Galileo preferred to teach his students how to use theories, instead of teaching them accepted knowledge.

Galileo's discussions with a number of scholars from the university at Pisa about floating bodies (1611-12) became notorious.[20] The Aristotelian scholars argued that a piece of ice, like any other object, floats or sinks dependent on its *shape*, and they sustained their argument by quotations from Aristotle's works. Based on experiments performed from his youth, Galileo showed the shape of a body to be irrelevant. Only its density determines whether it will be floating or sinking in water. He also mentioned an authority, namely Archimedes, but he sustained his argument by means of the hydrostatic balance. At the Grand Duke's request, Galileo published his views in the *Discourse on Bodies on or in Water* (1612). He made sure his argument was so clear that anybody able to read could understand it. Therefore he wrote this treatise in colloquial Italian.[21] The hydrostatic balance was easy and cheap to construct, and anybody could repeat Galileo's experiments.

We see that language, in order to show and clarify an argument, is not restricted to words. Experiments may also function to sustain an argument, to clarify it. Experiments in school physics have such a didactic function, besides other functions mentioned in Sec. 5.3.[22] A similar didactic function can be allotted to illustrations, graphs, *etc.*, and in particular to giving examples. Examples can never prove a theorem, but they can clarify it, and make it acceptable.

7.3. Polemic

Language not only has a didactic function in the relation between teacher and student, but also a polemic one, in the relation between scholars among themselves, especially if they disagree. In that case, the aim of polemic is to achieve support, sometimes by discrediting the aims of the adversary. Nowadays, polemic is somewhat less popular than in the 17th century, when all scientists practiced polemic passionately — at least in science, but in philosophy polemic is still in vogue.

Copernicus and Newton
Copernicus' *Revolutionibus* consists of six books. Books 2 through 6 contain a concise report of the mathematical theory of celestial motion on a heliostatic basis. These books being full of calculations are dull and difficult reading. The first book, including the Preface, contains a severely simplified summary of the theory, besides much propaganda. Copernicus does not hesitate to paint the state of affairs before the publication of his book as black as possible (see Sec. 4.5). He presents his theory as much simpler than it is, and uses completely irrelevant arguments, like the famous neo-Platonic confession: "In the middle of all is the seat of the Sun. For who in this most beautiful of temples would put his lamp in any other or better place than the one from which it can illuminate everything at the same time? Aptly indeed is he named by some the lantern of the universe, by others the mind, by others the ruler. Trismegistus called him the visible God, Sophocles' Electra the watcher over all things. Thus indeed the Sun as if seated on a royal throne governs his household of Stars as they circle around him."[23]

In his *Principia*, Newton especially opposed Cartesian physics.[24] The second part of his book had no other aim but to show that Descartes' theory of vortices is unfit, and the "General Scholium" at the end of the third book starts with the sentence: "The hypothesis of vortices is pressed with many difficulties."[25]

The polemical character of his book comes even to the fore in its very title. Cartesian physics is prominently displayed in Descartes' *Principia Philosophiae* (1644), Foundations of Philosophy. Descartes argued that physics should be studied in the mathematical way, *more geometrico*, in order to achieve equal certainty. Newton reproached him for not keeping his promise. Instead, Newton presented the genuine mathematical foundations of natural philosophy — *Philosophiae Naturalis Principia Mathematica*.

From their side, Descartes' disciples reproached Newton for introducing action at a distance, which they called an "occult" principle of explanation (see Sec. 3.5). Newton replied that the so-called "clear and distinct" ideas of Cartesian physics, like extension and hardness, are merely found by induction, no less than gravity.[26]

But the master of polemic was Galileo.

Galileo's polemic

Galileo used his language as an instrument of propaganda. His treatise on floating bodies, mentioned above, is not merely a learned discourse about a physical problem, but first of all it is a booklet aimed to ridicule his adversaries' method of argumentation. Also his *Letter to the Grand Duchess Christina*, his letters on the sunspots, his treatise on comets, and *Il Saggiatore* had a polemical rather than a theoretical aim.

The latter two especially belong to the category of polemical writings. In 1618, shortly after each other three comets appeared at the sky — and this was commonly considered to be a bad omen, an indication of approaching disaster — the thirty-year war, as it is now called, broke out. But Galileo was not concerned with astrology, but with the physical nature of the comets.

Tycho Brahe had conclusively demonstrated the comet of 1577 to move beyond the moon. Initially, Galileo agreed with this view, because it was applicable as an argument against the Aristotelian world view of incorruptible heavens and crystalline spheres. In this view, comets could only be sublunary, meteorological phenomena, but Tycho proved them to be celestial bodies, both corruptible and moving uninhibited by any crystalline sphere.

In 1618, however, Galileo no longer needed this argument. Not only the appearance of comets, but also that of novas (new stars) in 1572 and 1604 had convinced the astronomers of the untenability of Aristotle's theory of unchanging celestial spheres. Contrary to Galileo, who considered Tycho's system merely a variant of Ptolemy's, the Jesuit astronomers in particular considered it a useful compromise between Copernicus' and Ptolemy's systems. Against his intentions, Galileo had already collided with C. Scheiner, a member of the Jesuit order, about the priority of discovery and the interpretation of the sunspots.[27] The comets of 1618 became the occasion of a second collision with the Jesuits, who were his friends until 1616.

Galileo had a good reason to doubt Tycho's theory of the comets. Tycho had observed that if Copernicus were right about the annual motion of the earth, and if the comets move in circular paths around the sun, they should show retrograde motion like the planets do. The fact that the comets do not show retrogradations was interpreted by Tycho and the Jesuit astronomers as a proof against the motion of the earth. Only Newton was able to demonstrate that the absence of retrograde motion is due to the elongation of the elliptic orbits of the comets.[28] But Galileo could do little else but argue against their superlunary character.

About the three comets of 1618, the Jesuit Orazio Grassi wrote an anonymous treatise, *Disputatio astronomica de tribus cometis anni MDCXVIII*, in which he defended Tycho's views. Galileo's former pupil Mario Guiducci published a rejoinder, *Discorso della Comete*, written by Galileo, rejecting the view that comets are superlunary phenomena. About Tycho's proof he rightly observed that its validity depends on the assumption that a comet is a "thing" with a definite position and motion. In order to clarify this argument, we point to the rainbow, which is not a "thing." If we would try to determine the distance of a rainbow in Tycho's way, we would find it equal to the distance to the sun. Yet nobody doubts the rainbow to be an atmospheric phenomenon.

Grassi recognized Galileo hiding behind Guiducci, and wrote a defense under the pseudonym of Sarsi. He called it *Libra astronomica ac philosophica* (The astronomical and philosophical balance, 1620), as a reaction against Galileo's imputation that Grassi had wrongly assumed the comets to have appeared in the constellation Libra. In his turn, Grassi showed many of Galileo's arguments to be untenable.

Only in 1623 Galileo replied with *Il Saggiatore* (The gold balance). For the discussion on the comets it did not supply any new argument, but it is of polemical and philosophical interest, because Galileo used it to expound his views on scientific method. Because he did not hesitate to ridicule his opponent, after this episode Galileo lost all his friends among the Jesuits.[29]

Galileo's Dialogue and Discorsi

The acme of scientific polemic is Galileo's *Dialogue Concerning the Two Chief World Systems — Ptolemaic and Copernican* (1632). "Was the book an attempt to establish the truth of the earth's motion, or an attempt to induce or increase the acceptability of this idea in the mind of Galileo's contemporaries? If the former, the book was a failure, if the latter a success."[30]

Shortly after Galileo finished *Il Saggiatore*, his friend the Florentine Cardinal Maffeo Barberini became pope, and Galileo decided to dedicate the book to him. Pope Urban VIII was much charmed by the "gold balance", and during an audience in 1624 he advised Galileo to write a book in the form of a dialogue, in which Ptolemy's and Copernicus' systems would be discussed impartially. Galileo, who did not take this advice completely to heart, finished the book in 1630 and published it in 1632, after having overcome a lot of trouble with the censors.

The book is a dialogue between three persons during four days at Venice. The first day contains a criticism of the Aristotelian cosmology, in particular the distinction between heaven and earth. The second day is concerned with the daily, the third day with the annual motion of the earth. The fourth day is devoted largely to Galileo's theory of the tides, intended to prove the earth's motion (see Sec. 3.3).

The three characters involved are Salviati, who represents Galileo; Sagredo, an interested layman usually siding with Salviati; and Simplicio, named after the 6th-century Aristotelian philosopher Simplicius. He represents the conservative Aristotelians.[31] Sometimes, a fourth person is introduced, "the Academician", meaning Galileo himself, since 1610 a member of the Roman *Accademia dei Lincei*. Hence, it is three against one, not exactly impartial.[32] The name of Simplicio is easily associated with a simpleton, which he is not, however.[33] As a true Aristotelian he was not mathematically trained, and sometimes he defends an obviously untenable position.[34] Nevertheless, even though Simplicio always loses, his arguments are usually by no means unsound. Still, the choice of his name was to cause Galileo much harm, in part because it further alienated the Jesuits. In 1624, Galileo had agreed to include a conclusive and authoritative statement by the pope (see Sec. 9.2), who considered himself a progressive, broad-minded liberal. The task to put this statement forward Galileo assigned to Simplicio — the simpleton, the conservative, who always loses. The pope seems to have resented this to a high degree.

Galileo's tactic was first to strengthen his opponent's views as much as possible before refuting them.[35] But Galileo also omitted arguments if he knew he could not refute them. His most serious omission concerned the system of Tycho Brahe, which did not help to endear him to the Jesuits.[36] They could feel themselves cheated in a double sense. First, by the very title of the book. Galileo suggests that there are only two possible systems, Ptolemaic and Copernican, and by proving the first wrong, he justifies the second. This is an unfair use of the method of *"reductio ad absurdum"*, if there is a third alternative available, which the Jesuits took to be the Tychonian system. Many arguments against the Ptolemaic system do not apply to Tycho's.

Next, Galileo is not really interested in an *astronomical* problem, *i.e.*, the controversy between Ptolemaic and Copernican views, but in a *cosmological* one. The Ptolemaic and Copernican systems are only discussed with respect to their barest essentials (see Sec. 6.2). Galileo's criticism is not directed to Ptolemy's heterocentric system, but to Aristotle's homocentric cosmology (see Sec. 2.1).[37]

The *Dialogue* is running propaganda in favour of the Copernican creed of the moving earth, notwithstanding Galileo's protestations to the contrary. It is a justification of Copernicanism, *i.e.*, "..an attempt by verbal means and techniques to induce or increase adherence to Copernicanism."[38] But it is not only a polemical work, it is also very didactical. Galileo's explanation of the Copernican system and of retrograde motion is much more clear than Copernicus' own.[39]

All this is far less the case with Galileo's *Discorsi* (1638), which partially has the same dialogue form, starring the same characters. The *Discorsi* is mostly a scientific work, and it is less easily readable than the *Dialogue*. It is composed of Galileo's treatise *De Motu Locali* (On local motion), written in Latin before 1610, together with comments, written in Italian between 1633 and 1638. It is not concerned with astronomy or cosmology, but with mechanics, kinematics, and the properties of rigid bodies. It has little value as a polemic or didactic work, but as a theoretical treatise the *Discorsi* is probably more important than the *Dialogue*.

In the next chapter we shall have more to say about the relation between Galileo and the church. At present we may conclude that the prohibition of the *Dialogue* in 1633 was induced by its clarity, not by its scientific novelty. No other work has made as clear as Galileo's *Dialogue* that Aristotelian philosophy was defeated. Most subjects discussed in the book were already known to the specialists, but Galileo made them available to the general public. Moreover, Galileo made clear that the new philosophy of nature was at variance with a literal interpretation of some biblical passages, and he was therefore charged with heresy. This made his book a threat to the establishment, which hastened (but in vain) to withhold it.

8. Science and society

8.1. The organization of science

In this chapter we shall pay attention to the social aspect of science. During the Copernican revolution, Europe was in great turmoil with Reformation and Counter-reformation; religious, civil and colonial wars; the flourishing of superstitial practices like astrology, magic, and witch-hunting; periodic outbursts of the plague and of venereal disease; the decline of Italy, Spain, and the German Empire, and the rise of France, England, and the Dutch Republic; the emergence of bourgeois capitalism and overseas trade. It should be realized that the Copernican revolution in astronomy and physics was merely a marginal phenomenon, unnoticed by most contemporaries. We cannot but offer a few comments on the organization of science and its relation to the church, the only extant international organization after the decline of feudalism.

The decline of the universities
In the Greek and Hellenic periods, science as an activity was mostly concentrated in centres of education like Plato's *Academy* (387 B.C.-529 A.D.) and Aristotle's *Lyceum*, both at Athens. The *Museum* at Alexandria, having a world-famous library, was the most important centre of research between *c.*300 B.C. and 400 A.D.

During the middle ages the universities became the most important and almost exclusive centres of learning. The duplication and dispersion of manuscripts was usually taken care of in monasteries, where many hours were spend copying manuscripts, but this activity ceased rapidly after the invention of printing — perhaps the most important and timely contribution of artisans to the rise of modern science.

One of the most striking aspects of the Copernican revolution was its taking place outside the universities. Before the 17th century science was mostly studied at universities, dominated by

Franciscan and Dominican friars, who were usually conservative Aristotelians. The most important reform of education was enforced by Jesuits, who founded new schools and universities. For didactic reasons they composed epitomes of Aristotelian science, which came into use everywhere in Europe, even at the university of Leyden, which from the end of the 16th century gained a reputation as a reformed centre of learning. For quite a long time the Jesuits were considered to provide the best education available in Europe. Because most universities remained faithful to Aristotelianism, its opponents placed themselves outside the scientific community.

After his study in Italy, Copernicus became a canon at Frauenburg, which was a governmental rather than a scientific position. Tycho Brahe never held a university position. Kepler was at first a teacher at a minor school, but the most creative part of his life he served the Austrian emperor Rudolph II as Imperial Astronomer. Galileo was a university professor during an important part of his life, at Pisa and at Padua, but he only became an avowed Copernican after he was nominated Court Philosopher to the Grand Duke of Tuscany. Newton, too, after a long career as university professor, took his leave in order to live in London in a more rewarding function as officer of the Mint. Descartes, Pascal, Huygens, and Boyle never were connected to any university in a professional capacity.

Two new kinds of organization were to arise during the Copernican revolution — institutes and academies.

Institutes

The oldest scientific institutes were astronomical observatories, and one of the first European observatories was Tycho Brahe's. With financial support from the Danish king, who made the island Hveen (Venus) in the Sont available, Tycho built c.1580 two observatories, Uraniborg and Sternjeborg. He employed a number of assistents who carried out observations, experiments, and calculations. The astronomical instruments designed by Tycho were the best ever made before the invention of the telescope (see Sec. 6.5). In these observatories alchemy was studied as well as astronomy. After a conflict with the king, Tycho departed for Germany with all his instruments. In 1599 he became Imperial Astronomer at Prague, until his death in 1601.

For a long time, the astronomical observatory at Paris was dominated by J. D. Cassini, the leading astronomer of Europe between 1669 and 1712. He discovered four satellites of Saturn

(1671-1684), and his data were used by Newton. About 1676 Cassini was assisted by the Danish astronomer Rømer, who from observations of the periods of Jupiter's moons proved the speed of light to be finite.

A third example of an astronomical institute is the foundation of the Royal Observatory at Greenwich near London, where since 1675 without interruption regular observations have been made. The first Astronomer Royal was Flamsteed, his successor was Halley. Other well-known observatories dating from the 17th century are those at Rome (the Vatican) and at Leyden.

It is generally assumed that the professionalization of science only started in the 19th century. This is wrong as far as astronomy is concerned. Besides, several scientists were professionally employed as government advisors, and, of course, as university professors. It is true, however, that beginning with the 19th century, scientific activity became almost exclusively professional, with amateur science gradually fading out. But then the phenomenon of amateur science only flourished in the 17th and 18th centuries — neither before, nor after.

The academies

The academies to be founded during the 17th and 18th centuries in many countries were intended to promote scientific activity, to further the exchange of information, and to adapt scientific results to the public good. Almost always they had an outspoken anti-Aristotelian and anti-university attitude. Being progressive foundations, they did everything to disperse the modern view of science — "experimental" or "mechanical philosophy."

Galileo's membership since 1610 of the *Accademia dei Lincei* at Rome contributed much to his success.[1] The *Accademia* was headed by Federico Cesi and existed between 1603 and 1630, the year of Cesi's death. The *Royal Society for the Advancement of Science* was founded about 1660 by Boyle and others, and received its Royal Charter in 1662. Its first secretary, Henri Oldenburg, together with Robert Hooke, who from 1662 was curator in charge of experiments, did much to stimulate the Society's work. From 1703 until his death in 1727, Newton was its president. Finally, the *Académie des Sciences* at Paris, founded by Louis XIV in 1666 was initially directed by Huygens. These examples show that the Academies were largely in the hands of Copernicans.

It is one of the characteristics of Copernicanism to stress experiment and observation as sources of data (see Sec. 5.3). As a consequence, scientific work soon became more and more expensive.

This was one of the reasons for founding institutes and academies. Weak financial resources could be joined, and it became possible to ask the government to support scientific work.

In order to obtain this support, the academies stressed the social relevance of science. The usually slight efforts in this direction testify to the limited validity of this claim. The much improved methods of determining one's position at sea were more due to the craftmanship of instrument makers than to scientific insights. For a long time to come, astrology was considered to have more practical use (*e.g.*, in medicine) than astronomy.

8.2. The emancipation of science

The emancipation of science has a twofold character. The first has been mentioned in Sec. 4.5 — the emancipation from the authority of philosophy, from Aristotle's philosophy in particular, but also from Descartes'. Of course, it was not a liberation from any kind of authority, for modern science also recognizes its authorities. But these are now scientists, like Newton, Maxwell, Einstein, Bohr and Heisenberg, not philosophers or theologians.

The second kind of emancipation has a social character. After the Renaissance scientists have refused to accept the authority of any non-scientific organization in scientific matters.

We can be brief about the influence of the state. Neither in the middle ages nor in the 16th and 17th centuries were European governments concerned with science. Occasionally, a university was granted a privilege, the exclusive right to provide higher education, and in the 17th century governments started to promise and sometimes even to pay money to scientific academies. Nevertheless, before the 19th century the influence of the state on science was negligible.[2]

Internal and external historiography of science
For some time there has been a conflict between adherents of "internal" and "external" historiography of science.[3] The internalists consider history of science to be an autonomous process developing entirely apart from history in general. The externalists consider history of science to be part of social history. The internalists desire to take into account only the "rational", *i.e.*, logical elements of scientific developments. The externalists state that history of science is a product of the community, social and economic factors being highly relevant if not decisive.

It seems that among historians of science nowadays some kind of consensus is achieved, rejecting both extremes. It cannot be denied that many problems in science are provided by society, by extra-scientific needs. The social background of scientists is likely to be influential on what they consider to be satisfactory solutions of problems. On the other hand, we have seen that a good theory generates its own problems (Sec. 4.3). Therefore, science has its own internal dynamics, making it relatively autonomous with respect to social needs. In fact, often scientific research opens up possibilities which are transformed into social needs only afterwards.

The mutual influence of science and society is obvious in the 20th century, but was only marginal in the 17th century, when the most striking example was the problem of determining longitude at sea. Galileo thought that the periodic motion of Jupiter's moons could serve as a reliable clock, and he negotiated his idea with the governments of both Spain and the Dutch Republic. However, only after the invention of the ship's chronometer could this problem be solved, and this invention owes more to the crafts than to science. The same applies to the development of sextants, compasses, and similar instruments. Also the relation between science and warfare was sporadic during the Copernican revolution.

Marxist authors have attempted to prove that the Copernican revolution was caused by the emergence of bourgeois capitalism, but their arguments are hardly convincing.[4] Also the view that the rise of modern science was effected by the Reformation, or by the Puritan spirit, cannot bear the scrutiny of criticism.[5] This is not meant to say that either view is wrong. Rather it should be stressed that the relation between science and society is too complicated for us to arrive at such far-reaching and general conclusions.

The social background of Copernicanism
At least four social groups were involved in the rise of science during the 16th and 17th centuries.[6]
1. *The university professors*, the scholars communicating in Latin, were disciples of Aristotle, and represented medieval scholasticism. They were generally conservative friars, and opposed modern science. Exceptions were the Calvinist Ramus, who about 1560 tried to reform scientific education at the University of Paris, and of course Galileo and Newton, who were both rare laymen among their clerical colleagues. (Significantly, both were unmarried.)
2. *The humanists*, who were especially concerned with the reform of language and literature studies. They wished to purify the

ancient inheritance from the alleged medieval corruption. They were opposed to scholastic studies, hence anti-Aristotelian, but usually also anti-scientific. They contributed next to nothing to the development of natural science, except by weakening the authority of Aristotelian philosophy. They wrote in Latin and promoted the knowledge of Greek. A foremost example is Erasmus of Rotterdam.

3. *The artists-engineers* sprang from the emancipation of the arts, c.1500 (see Sec. 6.5). In the middle ages the craftmanship of artisans was developed to a high degree of sophistication, witness the cathedrals, the widespread use of water- and wind-mills, mining, and alchemy.[7] Science (including the "liberal arts") and the crafts developed completely apart, apparently oblivious of each other. The artists-engineers like Michelangelo, Leonardo da Vinci, and Tartaglia in Italy, Albrecht Dürer in Germany, Simon Stevin and Isaac Beeckman in the Netherlands, endeavoured to bring the crafts to a higher intellectual level. Working apart from the universities and writing in their native language, they contributed more than any other group to the rise of modern science. Tycho Brahe, Gilbert, Galileo, Descartes, Huygens, Newton, and many other Copernicans closely cooperated with artisans, and often were skilled instrument makers themselves. The merger of science and the arts after about 1500 was to be the most fruitful of all relations between science and society.

In France, the Netherlands and Great Britain, the artisans were often Calvinists or Puritans. Their emancipation has largely contributed to the rise of "experimental philosophy," which however can also be found in Italy, where Protestantism was never influential.

4. *The theologians* sometimes used the power of the church to influence the course of science. In the Catholic countries most university professors were friars, mostly Dominicans or Franciscans, whereas the Jesuits had their own colleges (see above). In astronomical matters, the conservative Dominicans usually held to Aristotelian views, whereas the Jesuits accepted Tycho's system. The Jesuit Clavius was the main architect of the calendar reform of 1582. Jesuit astronomers were the first to confirm Galileo's discoveries of 1609-12. But later Galileo came into conflict with the Jesuits like Christopher Scheiner (about the priority of discovery and the interpretation of the sunspots), Orazio Grassi (about the comets of 1618, see Sec. 7.3), and Cardinal Bellarmine (about the reality of the Copernican system, see below). It was generally believed that the Jesuit order, in particular Scheiner, was largely

responsible for Galileo's fate in 1633.

Although the theologians lost most of their influence on science after 1633, it should be observed that for a long time to come most if not all scientists considered theology far more important than science. The common view was that the study of natural laws was just another means to gain knowledge of God.[8] Contrary to the medieval scientists, their 17th-century protestant descendants considered themselves able to decide on questions of natural theology, and often theologians sought their advice on this subject.

8.3. Science and the church

Christian belief and the views inspired by it have exerted a large influence on medieval science, and on the rise and development of modern science, in particular with respect to the de-deification of nature, and the status of natural laws (see Sec. 12.2). In the present section we shall be concerned with the Bible as a possible source of data for scientific theories.

The church as an organized institute usually refrained from influencing science. Both in the middle ages and in the 16th and 17th centuries scientists often experienced greater freedom than their non-scientific contemporaries. Even during the great religious wars and the prosecution of witches, magicians and heretics, scientists were often free to move around. The Lutheran Rheticus was able to visit the Catholic canon Copernicus, whose work was published by the protestant Osiander, in Lutheran Nuremberg. The Lutheran Kepler was protected by the Jesuits in the predominantly Catholic environment of Graz, and he succeeded his fellow-Lutheran Tycho Brahe as Imperial Astronomer at the Catholic court of the Austrian emperor Rudolph II. The Catholic Descartes lived nearly twenty years in the protestant Dutch Republic, and the protestant Huygens played an important part in Catholic France. Only Ramus became a victim of the massacre of St. Bartholomew, 1572 at Paris, and Bruno was burned at the stake in 1600. Both events, however, had little to do with scientific convictions.

The Bible as a source of data

The toleration towards scientists in Catholic countries was not less apparent than in protestant ones. Copernicanism, too, was left in peace during more than 70 years after its birth. Some people have emphasized so-called anti-Copernican utterances by the church reformers, Luther, Melanchton, and Calvin, but this emphasis is exaggerated or even mistaken.[9] Luther only once made an oral statement about heliocentrism, during a table talk, at least if we may believe a report written 27 years after the event. Melanchton rejected the Copernican viewpoint because of his adherence to Aristotelian cosmology, but he protected Rheticus, the first Copernican after Copernicus. Melanchton's colleague Reinhold was to prepare the so-called *Prussian Tables* (1551) based on Copernicus' calculations (see Sec. 4.5). Calvin never said anything at all about Copernicanism.[10]

Afterwards, among protestant theologians one finds both adherents and opponents of the view that the earth moves, but in general they agreed that the Bible is not suited to provide scientific theories with data. Especially Calvin emphasized that the Holy Scripture is written for ordinary people and relates to directly observable states of affairs, to daily experience and common sense. For Calvin's disciples, this included the biblical view that the earth is fixed, and that the sun rises and sets. They accepted that this common sense view is not binding for scientific theories. Although not Calvinists, both Kepler and Galileo shared the opinion "...that the intention of the Holy Ghost is to teach us how one goes to heaven, not how heaven goes."[11]

Whereas the church reformers stressed the authority of the Holy Scripture and the personal responsibility of the faithful, Catholic theologians related the authority of the Bible to its exegesis by the church, in particular by the church fathers. After the Reformation, started by Luther in 1517, the Catholic church went through the Counter-reformation, in which especially the Jesuits played an important part.[12] In order to take the wind out of the protestants' sails, the Catholic theologians tended to ascribe to the Bible more authority than they did before, especially with respect to scientific problems. This early example of "fundamentalism" or "biblicism" partly determined the conflict of Galileo with the church.

Besides, since the 13th century, Aristotelian philosophy became part of the tradition accepted by the church (see Sec. 2.1). This was especially due to Thomas Aquinas, who wrought an uneasy kind of synthesis between Christian doctrines and Aristotelian theories. Because Galileo used Copernicanism to fight Aristotelian cosmology and physics, he inevitably came into conflict with the church. After his condemnation, Catholic theologians often considered Copernicanism a protestant heresy.

The Reformation held that each church member, whether theologian or layman, has the calling to read the Bible, and to take responsibility for his own beliefs. Catholic theologians held that the church and hence theology has the exclusive right to explain the Scriptures — for the benefit of the faithful, no doubt. Galileo was very imprudent when he endeavoured in 1615 to interpret the Bible on his own authority, and after his first encounter with the Inquisition in 1616, he carefully abstained from any personal exegesis of the Bible.

Galileo and the church
Until 1610 Galileo never openly declared himself in favour of Copernicanism. Even in his courses on astronomy he always put forward the traditional Ptolemaic theory. In 1610 he published his little book *Siderius Nuncius* (Message of the Stars), describing his first observations with the telescope.[13] After that Galileo was known as a Copernican. In the same year he moved from Padua to Florence, where he came into conflict with conservative philosophers, though not on the issue of the moving earth (see Sec. 7.3).

In 1612 the Neapolitan priest Foscarini wrote a pro-Copernican book, inducing some conservative Dominican scholars to wonder whether the hypothesis of the moving earth was contradicting the Bible. Shortly afterwards, the Jesuit Cardinal Bellarmine wrote a friendly letter to Galileo, expressing his opinion that the earth's motion might be discussed as a logical possibility in a scientific debate. But nobody should hold this hypothesis to be true, in view of the Bible and the teachings of the church fathers, as long as there was no conclusive proof of the earth's motion.[14]

In 1615, at the court of the Grand Duke of Tuscany, a discussion between several scholars took place about the relation of Copernicanism and the church's doctrines, and in particular about Galileo's views on this subject. Not having attended this discussion, Galileo found it necessary to respond by means of an open letter to the Grand Duchess Christina, the Grand Duke's mother.[15]

In this letter, Galileo states his opinion that the Bible does not intend to make statements about nature, but to relate religious truths. What the Bible incidentally says about nature is adapted to the comprehension of common readers, and therefore has less authority than statements based on sensory experience and reasoning. In fact, Galileo claims the priority of science over theology concerning the study of nature. He even suggests that in the case of conflict the theologians should carry the burden of proof — they have to prove the scientists wrong. Hence, Galileo came dangerously close to Calvin's views which parted with the medieval practice of "double truth" (see Sec. 2.1), which was still defended by Cardinal Bellarmine.

Unfortunately, Galileo weakened his position considerably, by giving his own interpretation of Joshua's miracle to halt the sun's motion in order to lengthen the day of the battle of Israel against the Amorites (Joshua, Chapter 10). Galileo argued that this passage can only be understood by accepting his theory that the earth's motion is caused by the sun's rotation. He implied that if the sun stands still ("in mid heaven") the earth's daily motion also ceases. This exegesis being hardly less miraculous than Joshua's action, it may be assumed that Galileo only included it to show how precarious it is to use the Bible as a source of data.[16]

Galileo's letter and Foscarini's book induced a relatively mild indictment by the Inquisition. Without mentioning Galileo, the Inquisition gave as its opinion — not as a binding conclusion — that the idea of the moving earth is absurd and formally heretical. Cardinal Bellarmine was instructed to serve Galileo a warning, which was executed orally. It is not known what exactly happened at this occasion.[17] An unsigned report survives, according to which Galileo was admonished not to teach the earth's motion, but Galileo never saw this report before his trial in 1633. Afterwards he received a letter by Bellarmine, confirming that Galileo was not condemned, but advising Galileo not to teach Copernicus' theory as being true. Galileo took this as permission to discuss it as a hypothesis, but also as an interdiction to connect Copernicanism with the Bible or the doctrines of the church, and he painfully stuck to this. In the *Dialogue*, Galileo only once refers to the Bible, in order to criticize an author who used biblical arguments in a "scandalous way."[18]

After this affair of 1616, Galileo kept quiet for some time. In the discussion on the comets he hid behind a former disciple, and moreover took pains not to defend the Copernican doctrine. Only in 1623 did he openly attack the Tychonic Jesuits in *Il Saggiatore*. The same year his friend Maffeo Barberini became Pope Urban VIII. He encouraged Galileo to compose a dialogue on the systems of Ptolemy and Copernicus (see Sec. 7.3).

Although Galileo pretended to be keeping his agreement to present in his book an impartial discussion of two equivalent theories, the *Dialogue* is unmistakenly plain propaganda for Copernicanism. Galileo openly declared his explanation of the tides to constitute a convincing physical proof of the double terrestrial motion (see Sec. 3.3).[19] The censor did not accept the original title of the book (*Dialogue on the Tides*) and changed it into *Dialogue concerning the two Chief World Systems, Ptolemaic and Copernican.*

Notwithstanding the censor's approbation, obtained with some difficulty, the book was prohibited immediately after the first copies reached Rome, and Galileo was summoned to the Inquisition. We shall be brief about the remarkable course of affairs, during the trial.[20] It is relevant to observe that the Inquisition had enough reasons to condemn Galileo, but remarkably few juridically valid reasons. After all, the book had passed the censor, and the backgrounds and contents of Galileo's first encounter with the Inquisition in 1616 were obscure. (Except for Galileo, all people involved were deceased). Three out of ten members of the court refused to sign the verdict. Probably it is only because of the pope, who was personally hurt and was ill-advised by the Jesuits, that Galileo received a relatively harsh punishment. It consisted of the public recantation of his Copernican views, the prohibition of his book, and lifelong imprisonment, soon to be changed into confinement to his own home.

Galileo never really revoked his views, as can be seen from the fact that he afterwards made a number of corrections to the *Dialogue*, but none of them to meet the objections of the Inquisition.[21]

Consequences

It is a bit exaggerated to consider Galileo a martyr of science. He was not tortured, he was not confined to a cell, not even during the lawsuit, he was treated with high regard, he was protected by persons at the highest level, including Cardinal Francesco Barberini (the pope's nephew), the Grand Duke and his ambassador at Rome.

To a large extent, Galileo himself was responsible for the events leading to his trial, because of his tactless attitude toward the Jesuits and the censor, and his dubious way of interpreting the agreement with the pope. But as we observed in Sec. 7.3, the conflict was almost inevitable because of Galileo's ability to present a clear and convincing case for the new and revolutionary world view.

For the development of science, Galileo's condemnation did not have many negative results. It induced Galileo to write his new science of mechanics, the *Discorsi* (1638), surreptitiously published at Leyden.[22] As a matter of course, protestants did not care about the views of the Inquisition, but even Catholic scientists often paid little attention to clerical statements. The consequences were severest for the Jesuits, the most faithful sons of the church. Up to 1824, when the ban on Copernicus' and Galileo's works was lifted, the Jesuits had to adhere to the Tychonian system. Another faithful son of the church was Descartes, who in 1634, learning of Galileo's verdict, decided to withdraw his book *Le Monde* from being published. But later on he acted as a Copernican in disguise, and the same applies to most of his Catholic colleagues.[23]

The consequences of Galileo's condemnation have been most damaging for the church itself. Since 1633 the church has the image of an enemy of science. Pascal, who was a Catholic, but an adversary of the Jesuits, wrote in his *Provincial Letters*: "In vain you have obtained a decree of Rome against Galileo, because this will not prove that the earth stands still. If reliable observations were available to prove that it turns around, then all people together could not prevent her from rotating, and could not prevent themselves from rotating with her."[24] In the 19th century this picture was strongly advanced, both by anticlerical currents, and by the clerical opposition against Darwinism, in France and in Great-Britain.

However, this image is utterly wrong and undeserved. Since the beginning of the middle ages the Christian church has more often promoted than opposed science, and one or two counterexamples are more than compensated by the predominantly positive attitude of the church with respect to science and learning. And we should not overlook that the most severe opposition to modern views originated from scientists and philosophers, not from the church, both in the case of the Copernican and that of the Darwinian revolution.

8.4. The social responsibility of scientists

We conclude this chapter with a few remarks on "social responsibility." It follows from the social norm, having two aspects. The first is "not giving offence" or "to pay respect", the second "to act for the benefit of all." The chief responsibility of a scientist is to publish his or her results. This is a consequence of the public character of science, but also has the effect of delimiting the scientist's responsibility.

The public character of science
If we assume the aim of science in a restricted sense to be the discovery and opening up of the lawfulness of reality, we find that

the main output of science is *knowledge of laws* — supposedly true statements about natural laws within a theoretical framework. If he finds them satisfactory, a scientist publishes his results. Literally, this means to make the results available to the public.[25]

Publication is the exclusive responsibility of the scientist. He decides whether he considers his results fit for publication. Usually other people may be able to veto his decision — his Ph.D. advisor, his boss, his employer, the editors of a journal, or a book publisher. In this case they share the responsibility with the scientist, who is nevertheless primarily responsible. This includes the mode of publication, and its level — specialist, popular, or educational (see Sec. 7.1).

The public character of scientifically achieved knowledge, its availability for anybody, rests on the recognition that natural laws are not privately owned. Natural laws are valid for everybody, and in principle anybody has a right to know the laws, as far as possible.

This principle has a practical consequence. If a scientist has found a law and published a law statement, he cannot be held responsible for anything which is done with it afterwards. In particular with respect to practical applications, no scientist is responsible for what other people decide to do, which is their own responsibility.

Galileo was aware of his responsibility to publish his results, even if it brought him into conflict with the church.

Yet it moves

Galileo's *Dialogue* (1632) is one of the pinnacles of the Copernican revolution, which was started and concluded by the publication of two other books, Copernicus' *Revolutionibus* (1543), and Newton's *Principia* (1687). Galileo's condemnation, on the contrary, was the lowest point in this history. As a matter of fact, after 1633 no more serious objections against Copernicanism have been raised.

As part of his punishment, Galileo was forced to recant his views. It is clear that he cannot be held responsible for this. The legend, according to which shortly after the verdict Galileo, striking the earth, would have said: *"Eppur si muove"* (and yet it moves) contains more truth than the enforced recantation of views which he had defended since 1610, without pause.

Galileo's responsibility was to see his views to the press.[26] After the trial of 1633, he concentrated on the writing and publishing of his final work, the *Discorsi* (1638). Its publisher wrote: "Since society is held together by the mutual services which men render one to another, and since to this end the arts and sciences have largely contributed, investigations in these fields have always been held in great esteem and have been highly regarded by our wise forefathers."[27]

The earth's motion is the central theme of the Copernican revolution, and Galileo was its chief ideologue. This concerns not only the *Dialogue* with its masterly discussion of all arguments pro and contra terrestrial motion, but also the *Discorsi*, in which Galileo laid the foundation of a new mechanics, such as was absent in Copernicus' *Revolutionibus*. The *Discorsi* manifests the new philosophy of nature, and presents most clearly the arguments against those used in denying the earth's motion. When Newton in 1687 completed the Copernican revolution he made use of mechanics built on the foundations laid by Galileo.

Giving offense

Whether Galileo was rightly convicted or not, there can be no doubt about one reason why he was prosecuted. He was charged because he had offended the pope (as representative of the church) and the Jesuit order, contravening the social norm not to give offence.

Descartes describes this norm as a "moral rule, derived from the method", namely, "...to obey the laws and customs of my country, constantly retaining the religion in which, by God's grace, I have been brought up since childhood, and in all other matters to follow the most moderate and least excessive opinions to be found in the practices of the more judicious part of the community in which I would live."[28]

In its most crude form, giving offense in an argument is called *argumentum ad hominem*, for instance, to question the validity of an argument suggesting that the person proposing it has a vested interest in its acceptance. However, the social norm of "not giving offense" should not be restricted to the sphere of insults. It also includes interests.

A scientist should be aware that the publication of his views and results of research can be influential, even dangerous. It means that a scientist is not only responsible for the decision to publish his results, but also for the mode of publishing. This responsibility may be shared by other scientists, in particular if it concerns communication on the level of education or popularization. It also means that scientists as a community should be concerned with the application of their results by other communities, industrial, military, or otherwise.

But since the church tried to suppress the publication of Copernican views, scientists have always demanded the right to decide for themselves whether to publish their results or not, and in
which way. It is significant that the Royal Society assumed the
right to grant the publication of scientific works. Thus, Newton's
Principia bears the *Imprimatur* of the Royal Society's *Praeses*,
Samuel Pepys.

The scientific church

It is a quite popular view nowadays that scientists ought to recognize their social responsibility. They ought to direct their research to problems which are "relevant" in a social sense. This
view has been countered by the observation that nobody can tell
whether or how fundamental research will be relevant in the
future. For instance, the social relevance of the investigation of
electricity during the 17th and 18th centuries was entirely absent,
but nobody can deny its ultimate fruits.

The norm of "social relevance" should not be obscured by the
empirically established fact that most scientists are not in the
least interested in the social relevance of their work. They do research because they are challenged by their problems. Problem
solving is the central dynamic motif of the individual scientist's
activity. That is the reason why it is easy for industries and
governments to employ scientists to solve their problems. Perhaps
this situation is just as well. A scientist does not carry   the
responsibility of deciding whether his results should be applied
in extra-scientific circumstances.

We emphasize that the conceptual development of the lawfulness of nature is a cultural goal in itself. It belongs to the vo-
cation of mankind to investigate the world. But nature is not given
to scientists, but to all people, who therefore have a right to learn
of the results of science, and have the responsibility to decide whether and how they will apply these results. The scientists should
not be burdened with this responsibility — they are not priests.
Nor should the scientific community, by hiding its conclusions,
withhold the responsibility of other people for their use of the
information. The scientist's responsibility is to acquire knowledge,
and to disperse it as widely as possible, by means of publication,
popularization, and education. This responsibility should never
restrict other people's responsibility. Therefore, we disapprove of
present-day scientologists who seem to strive after the
establishment of a "scientific church", which like the medieval
church has the pretension to know what is the best for all.[29]

9. Parsimony and harmony

9.1. The economics of science

After the lingual norm of clarity and the social norm of evading offence, we shall now be concerned with the economic norm of parsimony, and the aesthetic norm of harmony.

Ever since Ernst Mach wrote *The Science of Mechanics* (1883), the economic aspect of science has been a topic of discussion in the philosophy of science.[1] Mach states: "It is the object of science to replace, or *save*, experiences, by the reproduction and anticipation of facts in thought...This economical office of science, which fills its whole life, is apparent at first glance; and with its full recognition, all mysticism in science disappears."[2] "Science itself, therefore, may be regarded as a minimal problem, consisting of the completest possible presentment (*sic*) of facts, with the *least possible expenditure of thought*....The function of science, as we take it, is to replace experience."[3]

It cannot be denied that theories (rather than "science") have an economic aspect, as we shall see. But Mach's view that theories are characterized by the need to economize our experience already falters if we realize that a theory is more than just a set of statements. Theories are not intended to give a description of the world, but to predict, to explain, to solve problems, and to systematize our knowledge. This means that theories transcend mere description.

The economy of concepts and statements
Mach observes rightly that if we reconstruct the "facts" in our thought, we invariably restrict ourselves to those aspects which seem important to us.[4] This is even the case in our non-scientific, natural thought, and becomes more and more plain the more abstract our theories come to be. To give a name to a thing means to abstract it from its relations with its environment, and from changes which it may undergo in the course of time. Mach con-

siders a "thing" as a complex, a bundle of experiences, with a relative stability. "Properly speaking, the world is not composed of "things" as its elements, but of colors, tones, pressures, spaces, times, in short what we ordinarily call individual sensations. The whole operation is a mere affair of economy."[5]

In Mach's world view sensorial experience is predominant, whereas his view on theories is primarily economic. The economic process is assumed to start from simple, ordinary, familiar concepts, all statements or judgements being specifications of these concepts. A judgement says in how far a given thing deviates from a familiar one.

The economy of causal thought
Mach denies the existence of cause and effect in nature. "Nature is given only once" is Mach's favourite expression, and "equal effects in equal circumstances" never occur. It is only a matter of economy that we speak of cause and effect, deliberately neglecting the differences always occurring in actual cases.

Referring to Hume and Kant, Mach states that the idea of cause and effect only arises from the attempt to reconstruct facts in thought, and to relate various events. The experience that such relations can be found leads to the idea that they are necessary. Mach ascribes this idea to the existence of voluntary motions, and the changes we are able to produce in our environment. But he admits to have no answer to the question why the experience of causality instinctively develops itself, either in individuals, or in education. From our position, we suggest that humans have been created so as to be able to experience the (likewise created) causal structure of nature.

The assumption that the world has a causal structure apart from human experience may be considered a theoretical statement. We readily admit that this statement cannot be proved from our experience, but it seems to be tacitly accepted as a ground of any experiment. Also in experiments one attempts to abstract from "irrelevant" aspects, in order to investigate stable relations between variable parameters. Nevertheless, experiments are performed in order to have an objective check on one's thoughts. In particular, an experiment may have the function of checking whether a theoretically derived causal relation between different states of affairs really exists or is spurious. Such an experiment cannot ultimately prove the investigated causal relation, let alone the causal structure of the world, but it would be quite meaningless if the latter were not presupposed.

The economy of laws

According to Mach, laws are nothing but economic summaries of human experience. For an example he points to Snel's law of refraction. The table relating all possible values of the angle of incident and refracted light is exhaustively replaced by the simple formula sin a/sin b = n, n being the index of refraction. Mach comments: "In nature there is no *law* of refraction, only different cases of refraction. The law of refraction is a concise compendious rule devised by us for the mental reconstruction of a fact, and only for its reconstruction in part, that is, on its geometrical side."[6]

However, Snel's law transcends our experience in various ways. First, it is assumed to be valid for all kinds of materials, whether investigated or not. Next, it is supposed to be valid at all times and all places. Thirdly, it is supposed to be valid for any angle of incidence between zero and ninety degrees. The number of possible angles is infinite, but even our collective experience of these angles is finite. Fourthly, the law takes the angle or its sine to be a *real* variable in a mathematical sense, whereas in experiments the measured angles or their sines only have *rational* values.

Though a law statement is doubtless an economic summary of our experience, it is far more than that. Mach states that the transcendent character of laws is only accepted for the sake of economy. "We fill out the gaps in experience by the ideas that experience suggests."[7] "Thus, on the one hand, science must remain in the province of experience, but, on the other, must hasten beyond it, constantly expecting confirmation, constantly expecting the reverse. Where neither confirmation nor refutation is possible, science is not concerned. Science acts and acts only in the domain of *uncompleted* experience."[8]

The economy of mathematics

Mach considers the use of mathematics in the natural sciences to be an exclusively economic affair. Already the simplest operations of arithmetic have an economic sense. This is even more the case with the use of symbols like x and y in algebra. "Mathematics is the method of replacing in the most comprehensive and *economical* manner possible, *new* numerical operations by old ones done already with known results."[9]

Again it may be admitted that the use of mathematics has an economic aspect. But we have seen before that it seems to be far more reasonable to assume the world to have a mathematical structure independent of human thought, without denying that mathematical concepts, theorems and theories are man-made, no less than those of the natural sciences.

Mach's critique of atomism

Mach accepted the transcendental character of law statements for the sake of economy, if it is restricted to extrapolation and interpolation into domains inaccessible to direct experience. But he firmly rejected the introduction of theoretical entities like atoms, which cannot be found from the extrapolation of direct experience. Atomic theories were only acceptable if considered temporary and tentative, and should be replaced by other theories as soon as possible.[10] Mach rejected the kinetic theory of gases in favour of thermodynamics. The atomic theory was no more than a mathematical model comparable with spaces of more than three dimensions.

Hence, Mach would not agree with the view that theories deviate from common sense, explaining the observable from the unobservable, the unknown and hidden structure of the world (see Sec. 4.4).

Economy in theories

Though we reject its hypostatization, we agree with Mach that the economic aspect is a universal mode of human experience. This means that the norm of parsimony should also be applied to theories. It means that in a theory no more statements should be used than will be necessary for its purpose.

For instance, we have seen that a theory used to solve a problem is in need of a set of initial conditions or data. The principle of parsimony requires one to use as few data as possible. This is called "Ockham's razor": after a solution of a problem is found, erase as many special conditions as possible, in order to increase the strength of the solution, or the explanation, or the prediction.

9.2. The simplicity of the Copernican system

Following Mach, modern instrumentalists reject the view that a theory should be able to explain. They assume that a theory does not aim at truth, but only at prediction. It is sufficient if a theory is able to describe accurately coincidences observed by the senses. If confronted with two competing theories one has to prefer the most accurate one, and if both are equally accurate, the simplest one should be preferred.

In this section we propose to compare the Copernican and the Ptolemaic theories of planetary motion from this viewpoint. Both theories being equally inaccurate, Mach and his disciples argue that the only reason to prefer the Copernican system is its greater simplicity.[11]

Heliostatic versus geostatic theories

It is a widespread but mistaken idea that the only difference between the Copernican and the Ptolemaic system is that the first is heliostatic, the second geostatic. Impressed by Mach's relativism, reinforced by the success of Einstein's theory of relativity, many physicists and philosophers have maintained that the system of Copernicus is, after all, kinematically equivalent to that of Ptolemy.

The arbitrariness of the choice of either a heliostatic or a geostatic *viewpoint* or *point of reference* was fully realized by the Copernicans, who spoke of "optical relativity." In 1689, when visiting Rome, Leibniz tried to reconcile Copernicans and Tychonians by pointing to the relativity of space and motion. But Newton observed that the heliostatic viewpoint ought to be preferred because the sun is never far from the centre of mass of the solar system. It makes even less sense to distinguish heliocentrism from geocentrism (instead of heliostatic and geostatic viewpoints), for even in Copernicus' time it was clear that the distance between the sun and the earth is vanishingly small compared to the size of the universe.

But to give equal rights to Copernicus' and Ptolemy's theories from a kinematical point of view would mean a gross and even grievous underestimation of the historical course of affairs. For, to maintain that from a kinematic point of view the two theories are equivalent is only possible by accepting the principles of inertia and of kinetic relativity. But these principles have been found by the Copernicans in order to meet the objections against a moving earth. Those who rejected the Copernican system did so because they had physical reasons for rejecting the possibility of a moving earth. Hence, to state that the Ptolemaic system is equivalent to the Copernican system means to sever the former from its Aristotelian presuppositions.

However, the differences between the two systems are by no means exhausted by the one being geostatic, the other heliostatic. In the Copernican system, the planets move around the sun, in the Ptolemaic system, they move around the earth. Therefore, stating them to be kinematically equivalent means to confuse Ptolemy's system with Tycho Brahe's.

Copernicus' theory is not simpler than Ptolemy's

Not only Mach and his disciples, but also Copernicus himself stated that his theory was much simpler than the competing one, because it needed fewer circles. In his *Commentariolus* (*c*.1512) in which he had not yet fully developed his theory, Copernicus states that he needs only 34 circles.[12] Many authors have accepted this statement at face value, maintaining that Copernicus reduced the number of circles from 80 to 34, an impressive result.[13]

Unfortunately, it is not that beautiful. First, the most recent Ptolemaic calculations in Copernicus' time were based on the use of 40 circles, not 80. The number 80 occurs in the non-Ptolemaic and Aristotelian homocentric system of Copernicus' contemporary Fracastoro, whose system, however, did not have the accuracy obtained by the Ptolemaic or Copernican systems (see Sec. 2.1). Next, Copernicus ultimately used 48 circles, not 34.[14] Hence, the difference between the two rivals was much smaller than Copernicus suggested, and even to his disadvantage.

It can easily be seen that the transition from the geostatic to the heliostatic system properly gains no circles at all, and the transition from the earth to the sun as the centre of planetary motion yields at most a reduction of five circles, for each planet one, if the same accuracy is obtained. All considered, this is really not very much.[15]

By the way, Copernicus' theory of the motion of the moon is much simpler than Ptolemy's, but this has nothing to do with the Copernican credo of the moving earth. It only shows that Copernicus was an able mathematician.

Copernicus' theory is more complicated than Ptolemy's

In several respects the Copernican system is significantly more complicated than Ptolemy's.

1. By assuming that the earth moves, all results of observation, necessarily taken from the earth, are more difficult to interpret. Kepler had to spend a great deal of his time on determining the motion of Mars as "seen" from the sun[16] (see Sec. 3.5).

2. The description of retrograde motion is simpler from the Ptolemaic viewpoint than from the Copernican one.[17]

3. The assumption that the earth moves with a double motion causes many conceptual difficulties for anybody not accustomed to this counter-intuitive idea.[18]

4. Copernicanism made it necessary to enlarge the universe considerably in order to explain the unobservability of stellar parallax (see Sec. 2.4). It had to introduce an incomprehensibly im-

mense empty space between the spheres of Saturn and of the fixed stars.

Instrumentalists preferred the Ptolemaic system

Instrumentalism is by no means a modern invention. In Sec. 2.1 we saw that in the middle ages astronomy was generally interpreted in an instrumentalist sense. During the Copernican revolution, many peaceful, moderate people advised the Copernicans to accept a similar position.[19] Osiander, the Lutheran theologian who wrote an anonymous preface to Copernicus' *Revolutionibus*, adopted an instrumentalist view (see Sec. 11.1), and Cardinal Bellarmine advised Galileo to do the same (see Sec. 8.3). Pope Urban VIII forced Galileo to insert in his *Dialogue* the following basically instrumentalist declaration concerning Galileo's theory of the tides: "I do not therefore consider them true and conclusive; indeed, keeping always before my mind's eye a most solid doctrine that I once heard from a most eminent and learned person, and before which one must fall silent, I know that if asked whether God in His infinite power and wisdom could have conferred upon the watery element its observed reciprocating motion using some other means than moving its containing vessels, both of you would reply that He could have, and that He would have known how to do this in many ways which are unthinkable to our minds. From this I forthwith conclude that, this being so, it would be excessive boldness for anyone to limit and restrict the Divine power and wisdom to some particular fancy of his own."[20]

But Copernicus, Kepler, Galileo, and other Copernicans refused to accept instrumentalism (Sec. 3.3). They required a theory to be able to do more than predict; for instance, it should have explanatory power. But such a requirement implies a realistic view of theories.[21]

Hence, we sympathise with Popper saying that the 20th-century instrumentalists *betrayed* Galileo, who came into conflict with the Roman-Catholic church because of his refusal to behave as an instrumentalist.[22]

It should be observed that Tycho Brahe and his Jesuit disciples were also realists. In particular the Jesuit astronomer Clavius put forward realistic arguments in favour of Tycho's system.[23] In Rome, Clavius was one of the first astronomers to confirm Galileo's discoveries of 1609-10. After Galileo's trial, no competent astronomer, whether Tychonian or Copernican, adhered to Ptolemy's instrumentalist views.

9.3. Copernicus' and Newton's systems of the world

At first sight, both Copernicus' and Newton's systems of the world are basically simple, but a closer view shows this impression to be deceptive. Copernicus needed an additional set of epicycles and ex-centers to arrive at a fairly accurate description of the celestial motions, and Newton's theory became quite complicated as soon as applied to more than two bodies. Newton never found a satisfactory solution for the moon's motion. Actually, the "three-body problem" (earth, sun, and moon) turns out to have no analytic solution in Newton's theory.

Nevertheless, we shall argue that Newton's theory is more economical than that of Copernicus.

Copernicus' system
The basic axioms of Copernicus' theory state that the planets, including the earth but not the moon, circumvent the sun, moving uniformly in circular orbits. This axiomatic basis is not sufficient to provide an accurate description of the observed motions. Copernicus had to add epicyclic and excentric motions in order to achieve the same level of accuracy as his competitor.

Without any doubt this should be considered a fall-back into methods which he thought to have overcome. This might be the reason why Copernicus hesitated so long to publish his theory, about thirty years having elapsed between his *Commentariolus* (c.1512) and *Revolutionibus* (1543). It is why Kepler sought for a new theory, and why Galileo in his propaganda for Copernican-ism completely ignored Copernicus' use of Ptolemaic devices.

Quite apart from the question of how many circles Copernicus needed, the fact that he needed additional circles makes his system more complicated than he had hoped.

Newton's theory
After three centuries it is difficult to understand the impact made by Newton's idea of *universal* gravity. Besides mechanics and mathematics as presupposed theories, the basic axioms of Newton's system of the world are the law of gravity, and the assumption that gravity is the *only* significant force acting between celestial bodies, also acting on the surface of the earth and down to its centre. With respect to planetary motion, Newton takes pains to demonstrate that both friction and magnetism are negligible.[24] In a two-body system like the sun with one planet, his axioms allow one to solve the kinematic problem, in particular to show that

planetary motion satisfies Kepler's laws.

However, as soon as we consider a third body, not neglecting the force between the smaller ones, the theory may become horribly complicated, as Newton found to his dismay. He succeeded in predicting some small irregularities in the motion of the superior planets, but the solution of the problem of lunar motion eluded him. Concerning the calculation of the motion of the planets the Newtonian theory is even more complicated than that of Copernicus, who improved earlier calculations of the moon's motion considerably.

Still, we maintain that Newton's theory is basically simpler, first because the difficulties mentioned above could not be solved in any competing theory, and were ultimately solved in Newton's. Secondly, these difficulties were solved without taking recourse to old-fashioned ideas rejected in principle. As far as mathematics is concerned, this became possible because of the development of the calculus, for which Newton besides Leibniz laid the foundations. But Newton's theory is basically a physical theory, and the physical axioms, mentioned above, never needed additional hypotheses until they were corrected by Einstein's general theory of relativity.

Copernicus could never solve the problem of planetary motion from his basic axioms, and Kepler proved conclusively that this is intrinsically impossible. Newton's difficulties could be solved by remaining completely within the context of his theory. Though it is mathematically complicated, it is physically simple, being based on one single law of force, and being able to describe and explain planetary motion to a much higher level of accuracy than was possible with Copernicus' theory.

Kepler, Galileo, and Newton on parsimony
Besides Copernicus, Kepler, Galileo and Newton also used the argument of simplicity. They even ascribed the virtue of simplicity to Nature itself. Kepler said: "Nature likes simplicity and unity",[25] and against Ptolemy who ascribed a separate theory to each planet, he asserted: "Nature uses as little means as possible."[26]

According to Galileo, all philosophers grant that "...Nature does not multiply things unnecessarily; that she makes use of the easiest and simplest means for producing her effects; that she does nothing in vain, and the like."[27] He says: "...for ultimately one single true and primary cause must hold good for effects which are similar in kind."[28] Galileo argues in favour of his law of fall because motion at constant acceleration is the simplest possibility after uniform motion.[29] He also proves that the observed motion of the sunspots is much simpler to describe in Copernicus' system than in a geostatic one, and he considers this an argument in favour of the former. (Galileo's proof is right as far as the diurnal motion of the earth is concerned, but is less convincing with respect to the annual motion.[30])

Newton's views on parsimony are expressed in his "Rules of Reasoning in Philosophy", at the beginning of the third book of the *Principia*.[31] The first rule concerns theories: "We are to admit no more causes of natural things than such as are both true and sufficient to explain their appearances", but the added comment refers to nature: "To this purpose the philosophers say that Nature does nothing in vain, and more is in vain when less will serve; for Nature is pleased with simplicity, and affects not the pomp of superfluous causes."

Also the second rule is concerned with parsimony: "Therefore, to the same natural effects we must, as far as possible, assign the same causes."

Clearly, the Copernicans valued the principle of parsimony no less than the instrumentalists did.

The economy of the Copernican system
We have now presented arguments to show that Copernicus' system is less simple in some ways than either Ptolemy's or Newton's. Nevertheless, Copernicus' theory may be considered more economical than Ptolemy's, because it could achieve more by comparable means. In particular, it *explained* retrograde motion by a single principle. In Ptolemy's system it is by no means necessary that all planets show this phenomenon, and actually, the sun and the moon do not. Ptolemy introduced various circles for each planet separately, such that the motion of each planet is *described* (not explained) independent of the motion of the other planets, the sun and the moon included. But in Copernicus' system, the retrograde motion of all planets, and its absence in the case of the sun and the moon, are explained in one single stroke. Hence a celestial body not showing retrogradation cannot be a planet.

Next, the Copernican theory is more economical than Ptolemy's because it explains the coincidence of retrograde motion with opposition and maximum brightness, and because it enabled Copernicus to determine the order and the relative dimensions of the planetary orbits. The latter was impossible in Ptolemy's theory, and the presumed order of the planets could only be derived because of a weak and faulty argument.

9.4. The principle of harmony

Strongly related to the principle of parsimony, the aesthetic principle of harmony is another recurrent theme in the history of science. It appears that the aesthetic mode of human experience cannot be reduced to simplicity or to symmetry, though it is related to both. Pleasure, joy, delight, elation, concern a fundamental mode of experience, known by anybody, but impossible to define.

Inspired by the Pythagoreans, the Copernicans found harmony both in nature and in theories.

Copernicus' system
In the first book, in particular in Chapter 10, of his *Revolutionibus*, Copernicus became lyrical about his system. "In the middle of all is the seat of the Sun. For who in this most beautiful of temples would put his lamp in any other or better place than the one from which it can illuminate everything at the same time?...We find, then, in this arrangement the marvellous symmetry of the universe, and a sure linking together in harmony of the motion and size of the spheres, such as could be perceived in no other way."[32]

Copernicus' disciple Rheticus wrote: "But if anyone desires to look either to the principal end of astronomy and the order and harmony of the systems of the spheres or to ease and elegance and a complete explanation of the causes of the phenomena, by the assumption of no other hypotheses will he demonstrate the apparent motions of the remaining planets more neatly and correctly. For all these phenomena appear to be linked most nobly together, as by a golden chain; and each of the planets, by its position and order and every inequality of its motion bears witness that the earth moves..."[33]

Comparing this beautiful Copernican system with the clumsy machinery of Ptolemy, with its epicycles and deferents, equants and excenters, it is easy to forget that Copernicus needed these tricks just as much, after all. So, for the sake of justice, we should compare the obvious beauty of the sun-centred system with the no less obvious harmony of Aristotle's homocentric system. There is really no reason to reproach the Aristotelians for not being impressed by Copernicus' or Rheticus' lyrics. In particular, we should not overlook that Copernicus replaced Aristotle's harmony of *homocentric* spheres by a system that was truly *heterocentric* (having two centres, the earth, as the centre of the lunar orbit, and the sun), more so than Ptolemy's system which only introduced heterocentricity for the sake of calculations.

Kepler's harmony

In his *Mysterium Cosmographicum* (1597) Kepler thought to have revealed the secret harmony of the universe by his model of five symmetric polyhedra with their inscribed and circumscribed spheres (Sec. 2.5). Although he knew that the original model did not tally with the observed facts, he remained faithful to the idea throughout his life. When he discovered his planetary laws, he changed his spheres into shells, thick enough to accommodate the elliptical orbits. The sizes of the orbits and their varying speeds were related to musical harmonies.

Kepler's most mature work on harmony is his *Harmonice Mundi* (World Harmony, 1619). It consists of five books. The first two concern geometry and contain an extensive study of many kinds of polyhedra. The third book deals with musical harmony, the fourth with metaphysics, psychology, and astrology. The fifth book is mainly astronomical.

Harmonice Mundi is a remarkable hodge-podge of good mathematics, sound physics, and metaphysical speculation. The elliptic law is completely ignored, the area law is mentioned only once, and then in a faulty way.[34] The third law, also called the "harmonic law", now considered the most impressive fruit of Kepler's work, can only be found with difficulty.[35] This does not mean that Kepler underestimated its significance.

Apart from this law, the astronomical part of this work became the most famous because of the following exuberant passage: "The thing which dawned on me twenty-five years ago before I had yet discovered the five regular bodies between the heavenly orbits...; which sixteen years ago I proclaimed as the ultimate aim of all research; which caused me to devote the best years of my life to astronomical studies, to join Tycho Brahe and to choose Prague as my residence — that I have, with the aid of God, who set my enthusiasm on fire and stirred in me an irrepressible desire, who kept my life and intelligence alert, and also provided me with the remaining necessities through the generosity of two Emperors and the Estates of my land, Upper Austria — that I have now, after discharging my astronomical duties *ad satietatum*, at long last brought to light...Having perceived the first glimmer of dawn eighteen months ago, but only a few days ago the plain sun of a most wonderful vision — nothing shall now hold me back. Yes,

I give myself up to holy raving. I mockingly defy all mortals with this open confession: I have robbed the golden vessels of the Egyptians to make out of them a tabernacle for my God, far from the frontiers of Egypt. If you forgive me, I shall rejoice. If you are angry, I shall bear it. Behold, I have cast the dice, and I am writing a book either for my contemporaries, or for posterity. It is all the same to me. It may wait a hundred years for a reader, since God has also waited six thousand years for a witness..."[36]

If this reader was to be Newton, Kepler's discovery of the harmonic law had to wait nearly seventy years.

Galileo's obsession with circular motion

Although less exuberant than Kepler, Galileo also pointed repeatedly to harmony and beauty.[37] For him, the beauty of the Copernican system was expressed in uniform circular motion not confined to the heavens.

In the 14th century, the impetus theory was applied to circular motion, and some thinkers came close to the idea of inertia (see Sec. 3.1). Buridan, for instance, observed that a heavy millstone, having much impetus if turned around, is inclined to persist in its motion. For the same reason he suggested that the celestial spheres are not in need of a cause to continue their motion. Galileo agreed with this opinion.

In an ordered universe only finite and terminate motions, that is, uniform circular motions, do not disorder the parts of the universe.[38] This view prohibited Galileo from accepting Kepler's discovery of non-uniform, elliptic motion. If comets were celestial bodies, they would move in oval trajectories, putting an end to the supremacy of circular motion. This is another reason why Galileo assumed comets to be atmospheric phenomena[39] (see Sec. 7.3). Even the motion of animals he considered primarily circular.[40]

Galileo's concept of inertia concerned celestial bodies, moving uniformly in circles around the sun, or turning around their axis, as well as terrestrial objects moving frictionless on a horizontal plane.[41] In this case, "horizontal" points to circular motion, for the earth is spherical. Only circular motion can be uniform.[42]

In Sec. 3.3 we observed that Galileo recognized two fundamental or natural motions, uniform motion (at constant speed) and uniformly accelerated motion (at constant acceleration). Galileo connected both to circular motions in the following way.[43] On a horizontal plane he considered a number of objects starting simultaneously from the same point, moving at the same speed in various directions. At any instant, their positions constitute a circle of which the common starting point is the centre. Next he considered in one vertical plane a number of objects moving simultaneously on planes with various angles of inclination. If they start simultaneously from the same point, at any subsequent moment their positions lie on a circle having the horizontal line through the common starting point for a tangent. Galileo concluded: "The two kinds of motion occurring in nature give rise therefore to two infinite series of circles, at once resembling and differing from each other...(this constitutes)...a mystery related to the creation of the universe."[44]

Evidently, Galileo's universe was composed of circles.

Huygens

Huygens, too, speculated in a Pythagorean way about the cosmos. Like Kepler, he was convinced that the solar system could contain only six planets. After his discovery of Saturn's satellite Titan, he thought this completed the number of moons, it being the sixth after the earth's moon and the four moons of Jupiter, discovered by Galileo. Even in his *Cosmotheoros* (1698) Huygens speculated about a world harmony.[45]

Like Stevin, Kepler, and Galileo, Huygens studied music. Galileo's father Vincenzio Galilei came into conflict with Aristotelian philosophers about the theory of musical consonants, invented by the Pythagoreans. He argued that this theory was no longer in ac-cord with new musical practices. Both Stevin and Huygens tried to found a new theory, leading up to Huygens' remarkable division of the octave into 31 tones. So many Copernicans were interested in music, that Drake states that modern physical science is "the offspring of music wedded to mathematics."[46]

It was also Huygens who found a beautiful solution to the problem of impact between two equal bodies moving at different speeds, simply by first considering the same bodies colliding at equal speeds, and next transforming his solution to another reference system. Physicists have always found that the consideration of symmetry leads to beautiful solutions of problems. In modern times we find this in solid state physics as well as in sub-nuclear physics.

Such aesthetically appealing theories cannot fail to be very convincing. "The true and beautiful are one", Galileo said,[47] and many scientists share this opinion, even though harmony can never be the only argument for accepting a theory.

10. Criticism

10.1. Popper on criticism

In the present chapter we shall discuss some juridical aspects of
the use of theories: criticism and decision making. Scientific research is never an isolated activity. It consists partly of criticism
of earlier results, and in its turn, it should be submitted to criticism. Criticism and decision making are strongly connected; they
are nearly identical. Deliberate decisions can only be made after a
critical investigation of the possibilities available, and criticism
means to decide between accepting or rejecting scientific results.

Criticism and contradiction
Hence, it cannot be doubted that criticism is very important in science. In particular Popper has stressed its relevance to such an extent that his philosophy is characterized as "the critical approach."

The idea of criticism refers to the logical law of non-contradiction. Popper says "...contradictions are of the greatest importance in the history of thought — precisely as important as is criticism. For criticism invariably consists in pointing out some contradiction; either a contradiction within the theory criticized, or a
contradiction between the theory and another theory which we
have some reason to accept, or a contradiction between the theory
and certain facts — or more precisely, between the theory and certain statements of facts. Criticism can never do anything except
either point out some such contradiction, or perhaps, simply contradict the theory...criticism is, in a very important sense, the main
motive force of any intellectual development. Without contradiction, without criticism, there would be no rational motive for
changing our theories: there would be no rational progress."[1]

However much we agree with Popper, we cannot suppress the
feeling that he exaggerates the function of criticism. It cannot be
dispensed with, but Popper's emphasis on criticism easily leads to

the underestimation of other indispensable aspects of research, such as creativity, utility, commitment, and belief.

Popper on induction
First, we consider theories as instruments of criticism. Accepting a theory, we can test many statements about the world and subject them to critical analysis. In particular, theories are used to criticize data, based on hearsay, intuition, tradition, observation, experiment, or other theories (see Sec. 5.3). Popper ascribes the critical faculty of mankind exclusively to the use of theories, to deductive reasoning. In Sec. 6.1 we have seen that Popper is very critical of induction. Observing that theories are deductive schemes, and that induction cannot be justified by deduction, it is clear that no theoretical justification of induction can exist. Therefore, Popper's absolutism concerning theories causes him to reject inductive reasoning entirely.

Popper's overestimation of theoretical criticism stems from his distinction of "three worlds", the first world of physical states, the second world of mental states, and the third world of ideas, theories, arguments, and problem situations (see Sec. 1.2). The fundamental distinction between the first and second world is due to Descartes, whose mechanistic philosophy caused him to divide reality into *res extensa*, the physical world whose essence is extension identified with matter, and *res cogitans*, the mental world, whose essence is thought, the human mind.[2]

Like Descartes, Popper poses the question of how these worlds interact.[3] World 1 interacts with world 2, which in turn interacts with world 3. There is no direct interaction between worlds 1 and 3. Still, Popper's worlds constitute a hypostatization of the logical subject-object relation, with theories, statements, *etc.*, functioning as instruments. Popper does not recognize that everything in nature has many mutually irreducible aspects and relations, numerical, spatial, kinematic, physical (see Chapter 3), biological, sensory, logical, historical, and so on. Because everything has a logical aspect, everything can be fitted into the logical subject-object scheme. We can always tell whether something is a subject, an object, or an instrument of thought, even though this depends on the logical context. This explains (and only partly justifies) Popper's distinction of three worlds. But because the logical mode of experience is only one of many, Popper's distinction is not exhaustive, and if he says it is, he absolutizes the logical aspect of reality.

In particular Popper has no eye for the historical, cultural or

creative function of science, which is to open up the lawful structure of nature. Therefore he belittles science as far as it exceeds
theoretical critical reasoning. He rejects Kuhn's view on "normal
science." Induction is denied. Hypotheses can only be found by
trial and error. Observation and experiment are only relevant as
far as they help us to criticize theories — our sources of data are
mainly hearsay.

It is fascinating to see how the activity of a scientist is restricted, not to say downgraded, by Popper. Of course, nobody
wants to be called uncritical. Perhaps this is the reason why Popper's "critical approach" has raised so little protest. Why bother
about the so-called methods of science, if criticism is the only
thing that counts?

Popper on falsification and demarcation

Theories are not only instruments of criticism, they are also subject
to criticism. Theories should be testable and tested. Also in this
respect, Popper shows his dedication to deductivism.

From a deductivist point of view, a law statement can never
be verified in a logical sense, but it can be falsified. Because Popper hypostatizes theoretical deduction, he raises falsification to
be the main if not exclusive method of criticism. We have seen
that he had to moderate this view in order to meet some convincing objections (see Secs. 5.2, 6.1).

Because criticism is narrowed down to falsification, and science is restricted to theoretical reasoning, Popper easily arrives at
a new demarcation principle of empirical science.[4] It is a reaction
to the logical-empiricists' demarcation principle which stressed
the empirical *meaning* of theoretical terms.[5] A theoretical concept or statement has meaning only if it can be related to sensorily
observable states of affairs. Otherwise it is either tautological
(*i.e.*, mathematical or logical), or meaningless (in particular a
metaphysical statement or concept). Because "meaning" cannot be
falsified, Popper rejects this criterion. He considers a statement or
theory to be empirical, however, if and only if it can be refuted in
principle. Other statements are either logical (or mathematical),
or metaphysical — but metaphysical statements are not necessarily meaningless.[6]

In Sec. 5.2 we discussed Popper's view on falsification. The fundamental *logical* asymmetry between verification and falsification cannot be denied. An existential observation statement or datum can be verified, not falsified, whereas a universal statement,
in particular a law statement, cannot be verified by an observation

statement but is falsifiable in principle.[7] A universal statement can be falsified by an existential statement. This logical state of affairs is not diminished by the fact that any verification or refutation is tentative and liable to criticism.[8]

The difference in emphasis between the logical-empiricists (on verification) and Popper (on falsification) is partly reducible to their difference in emphasizing factual observation statements, and law statements. Popper overestimates falsification of universal statements and underestimates verification of data (which was overestimated by the logical-empiricists). Popper says that his principle of demarcation is only applicable to theoretical systems, because only these are testable.[9] But then the distinction between verifiable existential statements and falsifiable universal statements is irrelevant to the problem of demarcation, because theoretical systems contain both.

If we accept that existential statements can be verified, and give inductive weight to so-called basic statements (see Sec. 5.2), then Popper's falsifiability criterion for theories is useful and very important.

Popper on mathematics

It may be doubted whether Popper is very critical with respect to his view on mathematics. On the one hand he excludes mathematics from empirical science, sharing with the logical-empiricists the view that mathematics is part of logic. On the other hand, Popper does not hesitate to take an example from the theory of numbers in order to explain the distinction between verifiable existential statements and falsifiable universal statements.[10]

Popper works with two unproved theorems about prime numbers. The first is Goldbach's conjecture, stating that each even number larger than 2 can be written as the sum of two primes. It is falsifiable in principle, by giving one example of an even number not being the sum of two primes. The second theorem concerns twin primes, *i.e.*, primes differing by 2, such as 11 and 13, or 17 and 19. The theorem stating that the number of twins is infinite cannot be falsified.

Now Popper proceeds to show that the two theorems can be formulated in a single formula: "For every natural number x, there exists at least one natural number y such that the two numbers $x + y$, and $(2 + x) \pm y$, are both prime." The difference between the two theorems is given by the "$\pm$"-sign, the minus-sign concerning Goldbach's conjecture, the plus-sign the twin-theorem.

Popper puts this example forward in order to criticize the logical-empiricist view on the meaning of statements. Clearly, because both theorems can be expressed by nearly the same formula, they should be equally meaningful, yet one is verifiable, the other falsifiable in principle. It could be countered that also the distinction in relevance between verifiability and falsifiability cannot be that large, if the difference is only a plus- or minus-sign. From a logical point of view, the distinction between verifiable existential statements and falsifiable universal statements evaporates if one allows mixed universal-existential statements, like the above quoted formula.

But we first of all refer to this example to illustrate that at least some mathematical theorems are testable. Even Popper does not maintain that *every* statement in a theory should be independently refutable. In particular the high level axioms of either mathematical or physical theories are usually not open to direct empirical tests. In general, only low-level *consequences* of theories are directly testable. Popper's example leads to the conclusion that at least some theorems in the theory of numbers are "empirical" according to his demarcation principle of empirical science, meaning that mathematics is, at least partially, empirical, and not tautological.[11]

The reason why most people usually hesitate to consider mathematics "empirical" is that mathematics cannot be tested in the same way as the physical sciences, by observation and experiment.[12] They do not realize that this implies a restricted view of observation and experiment, only considering their *physical* aspects. It is small wonder that mathematical theories cannot be subjected to physically qualified tests. In Popper's case, the domination of his "first world" by physical states of affairs points to a similar kind of physicalism. Like Descartes, and like the logical-empiricists, Popper not only absolutizes the logical aspect, but also the physical aspect of reality. This physicalism prohibits the view that mathematics is an empirical science.

But mathematical theories can be tested as a matter of fact — not by physically determined observations and experiments, but by comparing their results with common sense experience, with the creative imagination of fellow-mathematicians, and in particular by comparing various mathematical theories with each other. If not, mathematical theories could not be criticized.

Paradoxes of confirmation
Popper makes use of the so-called paradoxes of confirmation to

criticize the possibility of corroborating theories by positive evidence.[13] The best known example is Hempel's paradox. The statement "all swans are white" may be confirmed by any observation of a white swan. The said statement is, however, logically equivalent to "Everything which is not white, is not a swan", which is confirmed by any object being neither white nor a swan. The observation of a yellow pencil is therefore a confirmation of the thesis "all swans are white."[14]

Being logicians, both Hempel and Popper accept the paradox as a valid argument. But Hempel accepts its consequence, and Popper rejects the possibility of confirmation because of its consequence.

One cannot escape the feeling that the paradox ls little more than a logical joke that can easily be evaded. The collection of all swans is much smaller than the collection of all things which are not white. We could easily agree to apply the principle of verification only to the smallest of the two collections, *i.e.*, to restrict the universe of discourse.[15] Perhaps it is not a simple matter to put this intuitively obvious solution into terms of formal logic, but then formal logic is of little use to scientific practice anyway. Therefore, this part of Popper's criticism of the method of verification does not seem to be very effective.

10.2. Immanent, transcendent, and transcendental critique

Ever since Kant, philosophers distinguish between immanent, transcendent, and transcendental critique. Immanent critique attacks a theory from within, by accepting its presuppositions, its axioms, and eventually its data, but nothing else. Its main method is to show the inconsistency of some of the theory's conclusions. Transcendent critique attacks a theory by confronting it with a competing theory.[16] In both cases, theoretical thought as such is not challenged. The use of theories as instruments of thought is accepted.

Transcendental critique directs itself to the use of theories and to theoretical thought as such. It questions the use of theories as instruments of thought as a matter of principle, it tries to explain why the use of theories leads to the growth of knowledge, and it investigates the methods of science.

The effectiveness of immanent and transcendent criticism
There is some difference of opinion concerning the effectiveness of immanent and transcendent criticism. Some people, restricting im-

manent criticism to a purely logical analysis of the internal consistency of theories consider it of little importance, because it is in general not difficult to keep a theory logically consistent. However, a less restrictive view on immanent criticism includes the investigation of the consistency of a theory's result with accepted facts or presuppositions. In that case immanent criticism is very important, but not necessarily decisive (see Sec. 5.2).

Scientists tend to accept contradictions between the results, or even the axioms of a theory, as exceptions or as anomalies, as long as no other theory is available, or if other theories are considered unsatisfactory. Thus, the "anomaly of water", the fact that water shrinks on heating from zero to four degrees Celsius, contradicted the axiom of the caloric theory of heat stating that any increase of temperature of a body is associated with its expansion. Nevertheless, the caloric theory of heat was generally accepted between c.1780 and c.1840.

Also a purely transcendent criticism is not very effective. Most people will not be convinced of the falsity of a theory merely by having pointed out an alternative. In the case of heat physics, about 1800 an alternative was available, the kinetic theory of heat, but it did not lead to the abandonment of the caloric theory. Only if immanent and transcendent criticism are applied simultaneously, may a fruitful debate arise, ultimately convincing people to reject one theory in favour of another one.

Copernicus' criticism of Ptolemy's system
Long before Copernicus entered the scene, astronomers had criticized Ptolemy's theory for its clumsiness. This can be admitted, even if one rejects Kuhn's view of a crisis preceding Copernicus' work (see Sec. 4.5). King Alfonso of Castille (13th century), the first Christian monarch to further astronomical research is said to have commented on the Ptolemaic system that he could have given some good advice if consulted on the day of the creation of the universe. The homocentric system of Eudoxus and Aristotle, though superior from a cosmological point of view, was never a serious rival to Ptolemy's as far as calculations were concerned, and was equally clumsy.

Copernicus' immanent criticism of Ptolemy's system was in no way original, and did not disturb anybody. But Copernicus delivered a very effective transcendent critique by presenting a new theory, showing that it was not inferior with respect to calculations, demonstrating it to be vastly superior because of its capacity to explain the coincidence of retrograde motion with some other phenomena, and proving it to be more fruitful by determining the size of the planetary orbits. It caused Tycho Brahe and other competent astronomers to change Ptolemy's system drastically, as we shall see in Chapter 11.

Because it simultaneously generated a number of new and grave problems, Copernicus' theory was rejected by the conservative majority. Only a few progressive scholars accepted the challenge of a new theory and new problems.

Galileo's critique of Aristotelian cosmology and kinematics

Galileo's critique of Aristotle's cosmology was both immanent and transcendent. In his *Dialogue*, Galileo criticized Aristotle's theory of the heavens, showing its inconsistencies and its lack of agreement with recently observed facts, like the moon's mountains, the sunspots, the phases of Venus, and Jupiter's moons. He argued extensively against the distinction between celestial and terrestrial physics, fundamental in Aristotle's cosmology.

But Galileo's immanent criticism could never have been effective if it was not reinforced by transcendent criticism, the proof that the alternative Copernican cosmology was confirmed by the observed facts that contradicted Aristotle's.

Similarly, because Aristotle's theory of motion had a strong basis in common sense experience, Galileo had to exert both immanent and transcendent critique on common sense knowledge. The immanent criticism started from common sense experience, like the insight that a ball moving on a frictionless horizontal plane will move forever. Galileo demonstrated that this piece of common sense contradicts Aristotle's view that violent motion ceases if no force is acting — which statement is also based on daily experience.

Ultimately, he shows that common sense experience should not be the start of an explanation, but its end. Theories should be able to explain common sense, and Galileo does so by a theory starting from the law of inertia, a counter-intuitive law. This is the transcendent phase of his criticism, without which the immanent phase could not be effective. But conversely, the new theory would never have been convincing, if it were not preceded by the immanent critique of Aristotle's theory, showing that it was not as obvious as it seemed to be.

Newton's criticism of Cartesian physics

Newton's critique of Cartesianism was displayed on various levels. The most fundamental level concerns the principles of "mechanical philosophy." The Cartesians held the primary properties of matter to be found as clear and distinct evident ideas. These include the identification of matter and space, the rejection of the void, the law of conservation of motion, action by contact, and mechanical concepts like impenetrability, and hardness. Newton's immanent criticism consisted of showing that these Cartesian ideas are by no means evident, and are actually found by extrapolation of common sense experience. His transcendent criticism consisted of the introduction of action at a distance, the vacuum, and his own laws of motion. This criticism is transcendental as far as Newton (like Galileo) rejects the methodological principle that the starting points of theories should be intuitively clear.

The next level concerns method. Newton teaches his contemporaries how to apply mathematics to physical problems, implicitly defaulting Descartes' so-called mathematical method as being too superficial. Next, Newton teaches how to use experiments. He investigates mathematically the consequences of the hypothesis that the resistance in a fluid is either proportional to the speed of a moving body, or to its square. Then he discusses experiments, in order to find which hypothesis is actually true. He even designs an experiment to examine Descartes' claim of the existence of "a certain ethereal medium extremely rare and subtile, which freely pervades the pores of all bodies."[17] He finds it to be wrong, because it would imply that the resistance of a pendulum should be proportional to its volume instead of its cross section.

The third level concerns the theories of planetary motion. Newton's immanent criticism shows Descartes' theory to be inconsistent. It cannot explain why the planets move without loss of motion (i.e., frictionless) in a plenum. It cannot explain why the planets move according to Kepler's laws.[18] His transcendent critique shows that the alternative theory is superior. By the assumption that interplanetary space is void, Newton explains why planetary motion is able to go on for ever. From his law he explains why the planets move according to Kepler's laws. Moreover, his theory generates and solves many more problems than Descartes'.

It should be observed that Newton, in order to criticize Descartes' theory of planetary motion, first developed his own theory of motion in a resistive medium.[19] This theory does not play a constitutive part in his theory of planetary motion, but was necessary to show the Cartesian cosmology to be wanting.

Newton concludes the second book of *Principia* by: "...so that the hypothesis of vortices is utterly irreconcilable with astronomical phenomena, and rather serves to perplex than explain the heavenly motions. How these motions are performed in free spaces without vortices, may be understood by the first Book; and I shall now more fully treat of it in the following Book."[20]

In the General Scholium at the end of *Principia*, Newton repeats: "The hypothesis of vortices is pressed with many difficulties."[21]

Transcendental critique

Immanent criticism *of* a theory accepts its axioms but investigates its internal consistency, and its consistency with accepted data and presuppositions. Criticism *by* a theory means the examination of data, and their sources — in particular, common sense, observation and experiment. Transcendent criticism is the confrontation of two rivalling theories, by showing one to be superior to the other. It challenges the axioms of the theory to be scrutinized. Both immanent and transcendent criticism concern the logical and pre-logical aspects of theories: contradiction, consistency, meaning analysis of concepts, the truth of statements, and the predictive, explanatory, problem solving and systematizing powers of theories.

On the other hand, transcendental critique transcends logic. It concerns first of all the heuristic of science — how theories are found. In Chapter 6 we criticized Popper's method of "trial and error", "conjectures and refutations", not because it is entirely wrong, but because it is much too narrow. Lakatos elaborated the idea of "research programmes", which he drew from Popper,[22] into his "methodology of scientific research programmes' (Sec. 6.2). We have attempted to show that this is only one method, and we developed a "theory of the scientific opening process" (Sec. 6.6). Whatever its worth, it is a *transcendental* theory. There is nothing contradictory in proposing a *theory of theories*. It underlines the universal character of the logical aspect of human experience, meaning that it can be used to study the post-logical aspects.

Therefore, transcendental criticism also concerns the lingual, social, economic, aesthetic and juridical aspects of the use of theories, discussed in Chapters 7-10, and the aspects of commitment and belief, to be treated in Chapters 11 and 12. In order to be able to exert this criticism, we are in need of theories transcending the theories we want to criticize. For instance, we have to know what "clarity" is, and how it functions in scientific debates, in education, and so on, before we are able to criticize the lingual presentation of Newton's theory of gravity. We have to know what the "economy of science" means, before we can criticize Mach's hypostatization of the economic aspect of theories.

This kind of criticism is often called *metaphysical* (or metamathematical, or metatheoretical), or *philosophical*, but this should not give the impression that scientists do or should keep aloof from it. A great deal of scientific activity is making decisions, requiring all three kinds of criticism. It is possible and meaningful to distinguish immanent, transcendent and transcendental criticism, but it is hardly ever possible to separate them.

The highest level of transcendental critique, the critique of theoretical thought itself, will be deferred to Chapter 12.

10.3. The critical function of the scientific community

Criticism is subject to the principle of justice, in particular if decisions have to be made concerning theories. During the development, assessment, publication, and application of theories many decisions have to be made which are at least partly subjective. It has to be decided whether the data are reliable, whether the problems are relevant, whether the solution of a problem is satisfactory, whether the evidence of an empirical generalization or correlation is sufficient, and which heuristic will be applied to achieve one's end. Decision making has both an objective and a subjective side.[23]

Judgement by peers
In these decisions the scientist (or group of scientists) working on a problem is primarily responsible. However, sooner or later he has to subject his decisions to the judgements of others — his supervisor, the director of the laboratory, subsidizing agencies, and the editors or referees of a journal in which he wants to publish his results.

Since the emancipation of science from philosophy and theology, a scientist is supposed to subject his decisions only to the judgement of his peers. Only scientists are supposed to be competent to criticize scientific results. An important motive to publish one's results is the ambition to be recognized by one's peers. Criticism from other sides is not welcome, and usually ignored. This is not the case, of course, with respect to the question whether and how scientific results are applied outside science. Also, in particular in our century, the decision as to which kind of research will be stimulated is to a large extent made by non-scientific agents, like governments.

Impartiality

By the establishment of academies and scientific journals (Secs. 7.1, 8.1), Copernicanism laid the foundation of an internal scientific system of objective, impartial judgement based on some kind of common law. Being impartial or unbiased is a norm for juridical action, a norm which is valid, even though everybody knows that it can never be completely satisfied.

In particular the editors and referees of journals have a duty to pass a competent and impartial judgement in order to warrant the objectivity of the contents of the papers. A good editorial policy guarantees the reliability and credibility of the published theoretical and experimental results. The most prestigious journals are those which meet this requirement at the highest level.[24]

Criticism concerning the publication of scientific results is of great importance in science. The reader of a scientific paper does not need to rely on his own restricted insight, or on the authority of the often unknown author, but he may rely on the critical attitude of the editors of the journal, which may or may not have a good reputation.

Individual and communal responsibility

A scientific paper is signed by a scientist (or a few scientists) who is (are) responsible for its content. Usually, his name is accompanied by the name and address of the laboratory or institute where he is employed, meaning that it shares his responsibility. In the Western world, nobody else will be held responsible for the contents of a scientific paper, not even the editors of the journal in which it is published.

The scientific world is only partly organized into national and international societies, which often operate independently of each other. Their task is not to decide on the contents of papers or books, but to stimulate scientific communication by editing journals, organizing conferences, *etc.* Each scientist is free to join some and ignore other societies.

The organization and discipline of scientists is more advanced in the USA than in Europe, and even more in the USSR, where all science is conducted under the supreme responsibility of the Academy of Sciences, which is directly responsible to the Communist Party's Central Committee. Individual academic freedom, therefore, is probably larger in Europe than in the USA and the USSR.

Science and theology

We have seen that Copernicus rejected criticism from incompetent people by his statement: "Mathematics is written for mathematicians." The full statement reads: "There may be triflers who though wholly ignorant of mathematics nevertheless abrogate the right to make judgements about it because of some passage in Scripture wrongly twisted to their purpose, and will dare to criticise and censure this undertaking of mine. I waste no time on them, and indeed I despise their judgement as thoughtless. For it is well-known that Lactantius, a distinguished writer in other ways but no mathematician, speaks very childishly about the shape of the Earth when he makes fun of those who reported that it has the shape of a globe. Mathematics is written for mathematicians, to whom this work of mine, if my judgement does not deceive me, will seem to be of value to the ecclesiastical Commonwealth over which your Holiness now holds dominion."[25]

To be sure, Copernicus accepted the judgement of the pope, but he significantly did so by including him into the fraternity of scientists: "However, what I have accomplished in that respect I leave to the judgement of your Holiness in particular and of all other learned mathematicians..."[26]

Whereas Copernicus was prudent enough to include the pope among the competent judges of his science, Galileo outright rejected the competence of theologians (see Sec. 8.3). He could not evade the official censure of the church when he published his *Dialogue* (1632), but his *Discorsi* (1638) was published at Leyden, where no preventive censure existed.

After Galileo's trial, common practice among Catholics and protestants alike became to separate theology from science. The Royal Society even made this statutory, by excluding theology from its discussions. Although in England many scientists were divines, they usually agreed to keep their theological and scientific activities apart. The Royal Society considered itself the supreme court of justice in matters scientific. As we observed in Sec. 8.4, it is highly significant that without asking for ecclesiastical or governmental assent, it assumed the right to censure scientific publications, as can be seen from the title page of Newton's *Principia*, which bears the *Imprimatur* of the President of the Royal Society.

Newton

In the 16th and 17th centuries publication of one's scientific results was by no means taken for granted like it is nowadays. Copernicus hesitated a very long time to publish his *Revolutionibus*, probably because he feared the judgement of his peers.[27] (There was no reason to fear the censure of the church.) Descartes postponed the publication of *La Géometrie*, written about 1619 to 1637. Beeckman never published anything, and the relevance and originality of his insights only became clear after the publication of his diary in the 20th century.

Snel's work on optics, containing the discovery of the law of refraction, was never published and his manuscript is lost. Pascal valued his scientific work lower than his theological disputes, and after his death his inheritors hesitated quite a long time before they decided to publish his masterwork on aero- and hydrostatics, fearing that it would damage Pascal's reputation. Huygens' works were usually published with great delay, some even posthumously.

The most famous non-publisher was Newton, who until the end of the 17th century hated publicity. In particular his mathematical discoveries were never properly published, which led him into conflicts about priority with several people. He delayed the publication of his optical experiments, made about 1667, to 1672 and 1675.[28] Because of a conflict with Hooke and critical comments by others, he decided not to publish again on this subject as long as Hooke was alive. Hooke died in 1703, and Newton's *Opticks* was published in 1704, but the bulk of the *Opticks* is literally the same as a manuscript dating from 1672.

Meanwhile, at least up to 1685, Newton was more concerned with alchemy and theology than with mathematics and optics. Newton's attitude towards publication of his results resembles that of Copernicus. In mathematics, Newton had no other peers than Huygens initially, and Leibniz later on. Apparently, he did not like to invite criticism of incompetent mathematicians. Also his dealing with his optical discoveries point in the same direction. In 1672, Newton made available to the Royal Society his invention of the reflecting telescope, and his experiments with prisms on the spectrum of light. Newton was convinced that his arguments, derived from experiments, proved conclusively the heterogeneity of white light.

Initially, his publications were favourably met, but soon critical objections were raised by Huygens, Hooke, and others. Newton disproved the most reasoned arguments, but he became angry with several unwarranted statements put forward by Hooke and others. Although Hooke considered himself the outstanding expert on the science of colours, and Newton admitted to have learned a few things from Hooke's *Micrographia* (1665), he recognized Hooke's objections against his views to be wide of the mark and incompetent.

The publication of Newton's *Principia* (1687) occasioned another conflict with Hooke, and as we have seen Newton decided to abstain from a popular representation of his "System of the World" (Sec. 7.1), in order to avoid incompetent criticism.

Priority disputes

Science is a creative activity, meaning that priority of discovery is considered important. A pressing reason to publish one's results is to establish one's achievements, and in particular one's priority. During the Copernican revolution this was as important as it is now.

To protect one's priority without revealing a discovery, 17th-century scientists used to send each other an *anagram*, an unsolvable puzzle which afterwards could be used to claim a discovery. Thus, in 1610 Galileo sent Kepler the anagram

SMAISMRMILMEPOETALEUMIBUNENUGTTAURIAS,

meaning: *Altissimum planetam tergeminum observari* (I have observed the highest planet in triplet form), but Kepler interpreted it as: *Salve umbisteneum geminatum Martia proles* (Hail, burning twin, offspring of Mars, *i.e.*, Mars has two moons), complaining that this was barbarous Latin.[29]

Galileo's dispute with Scheiner on the discovery and explanation of the sunspots was mentioned in Sec. 7.3. Other Copernicans became involved in conflicts over priority, the most famous case being Newton's dispute with Leibniz concerning the discovery of the calculus. Present day historians agree that both were independent, and Leibniz was the first to publish the theory. But Newton suspected Leibniz of having read a manuscript by Newton before Leibniz' publication. It is clear that Newton was most to blame in this affair, not only because of his false accusations, but first of all because he failed to publish his results at the right time.

The objective grounds for criticism

Just as in court, in order to do justice it is not sufficient to have impartial and competent judges. They also need objective evidence. In the case of scientific research, this evidence is based on common sense experience, imagination, observation, measurement, and experiment — in short, the sources of knowledge discussed in Sec. 5.3. It is also based on theories insofar as they are accepted by the judges, and not challenged in the paper to be judged. If a new theory is presented together with experimental results, the judges have to consider the consistency of the two, and their reliability.

Both inductivists and deductivists often give the impression that scientific judgement, either by the scientist himself, or by external judges, should be a completely objective affair. It never is. Decision-making is a subjective art, based on objective evidence that has to be weighed, to be evaluated. Neither in courts nor in scientific affairs can any decision be made completely objectively. This is not something to be deplored. It keeps science a human activity, as it should be. And it opens the possibility for scientists to be committed to theories, as we shall see in Chapter 11.

Astrology
The scientific community also decides what kind of evidence will be admitted. Since the 17th century, theological as well as astrological arguments are excluded from the physical sciences.

It is only partly justifiable to charge astrology with superstition, subjectivity, lack of criticism, inconsistency, and unreliability. Far more important seems to be the fact that astrology has no relation whatsoever with astronomy, physics or biology. Astrology does not contribute to the development of these sciences, and conversely, it depends on them very little, if at all. Modern scientists do not show astrology to be wrong, but instead show that it lacks proof.

This state of affairs was not clear in the 16th century. Astrology fitted very well into the Aristotelian cosmology, because its main axiom is the correspondence between macrocosmos and microcosmos, the organic unity of the cosmos. As far as we know, Copernicus kept aloof from astrology, but Tycho Brahe, Kepler and Galileo practiced it. However, the emerging philosophy of mechanism eventually broke entirely away from astrology. In this respect science has exerted surprisingly little influence. In our day astrology remains as prominent as ever, despite the antagonism of scientists.

11. Commitment

11.1. "On the hypotheses in this work"

This chapter deals with the ethical principle of commitment. A reliable scientist should hold to his theories, defend his assumptions, be honest with his data, be fair in reporting his experiments, give credit where it is due — in short, behave such that he can be trusted with the development and application of his theories. Apparently, charity and criticism are contrary attitudes, but actually both are needed, in science as elsewhere.[1] Charity in reasoning means that one considers the arguments of one's opponent in a favourable light, and as seriously as he deserves. Hence, charity without criticism is as ineffective as criticism without charity, that is, criticism without mutual respect. We shall focus our discussion of commitment on the distinction between "hypotheses" and "axioms."

In Sec. 4.1 we distinguished axioms, presuppositions, and hypotheses. Nowadays this distinction is less fashionable than it was during the Copernican revolution.[2] In particular instrumentalists and adherents to the hypothetical-deductive method take all universal statements to be hypotheses. Although "hypothesis" means "supposition", we reserve the term "presupposition" for law statements borrowed from alternative theories and accepted as true. For instance, mathematical theorems are usually important presuppositions for physical theories. Axioms, likewise taken to be true, are the law statements characterizing a theory. Hypotheses, on the other hand, may or may not be considered true, and may or may not be law statements. They have a tentative character, only introduced to investigate their fruitfulness, or to test the ability of the theory to solve problems.

Osiander
The preface to Copernicus' *Revolutionibus* called *On the hypotheses in this work* was written by Osiander, the Lutheran minis-

ter who saw Copernicus' book through the press.[3] In this anonymous preface, Osiander adopted an instrumentalist view. He says:
"Nor is it necessary that these hypotheses should be true, nor indeed even probable, but it is sufficient if they merely produce calculations which agree with the observations." In the present section, we shall examine this claim.

Copernicus most probably did not share it, but he was re-
luctant to say so. In his own preface to Book I, Copernicus spoke
about "principles and assumptions, which the Greek call 'hypotheses.'"[4] In his *Commentariolus* (c.1512), fully entitled "Nicholas Copernicus, Sketch of his Hypotheses for the Heavenly Motions", he wrote about hypotheses as "statements called axioms."[5]
Contrary to Osiander, Copernicus did not distinguish axioms from
hypotheses. We shall return to Copernicus later on.

As early as 1569 the French professor Petrus Ramus protested
against the anonymous preface, which he incorrectly ascribed to
Rheticus. Ramus offered his chair at the Paris university to anyone who could design an astronomy free of hypotheses. In 1597 and
1609, Kepler claimed this chair, jokingly, for Ramus was dead by
then.[6]

Kepler on hypotheses

Kepler unmasked Osiander as the author of the anonymous preface to the *Revolutionibus*, and expressed his indignation at its
message. According to Kepler, Osiander's view contradicted Copernicus', because Copernicus believed his system to be true.[7] Kepler emphasized Copernicus' superiority over Ptolemy, because the
new system does not merely describe celestial motions, but also explains some of the associated phenomena (Sec. 2.4).

However, Kepler soon found himself in an awkward position.
Though greatly admiring Copernicus' theory, he was expected to
defend Tycho Brahe's system as a courtesy in return for using Tycho's extensive data. Moreover, on the basis of his calculations concerning the motion of Mars, he had to conclude that both Copernicus and Tycho were wrong. Kepler's *Astronomia Nova* (1609)
investigates four alternative theories: Ptolemy's, Copernicus', Tycho's, and ultimately his own.

In order to keep his promise to defend Tycho against the
claims of Nicolas Reymers Bär (*Bär* = Bear = *Ursus* in Latin),
Kepler composed *Apologia Tychonis contra Ursum* (1600-01),
which, however, he never published. Bär was the predecessor of
Tycho Brahe as imperial mathematician (astronomer) to the court
of Rudolph II at Prague, and Kepler was Tycho's successor. Four

years after Ursus visited Tycho (in 1584), he published a system which later turned out to be Tycho's. Tycho claimed priority, although the idea that the sun travels around the earth, and the planets (at least Mercury and Venus) turn around the sun, was already known in antiquity as the "Egyptian system."

In the *Apologia*, Kepler rejects the then-prevalent view, also defended by Ursus, that astronomical hypotheses need not be true; this contrasts them with the axioms of Aristotelian physics, which should not only be true, but be seen to be true by directly available evident insight (see Sec. 4.4). Insofar as Ptolemy's and Brahe's theories give good results, Kepler maintains that despite the falsity of many of their hypotheses, this is due to a portion of the hypotheses corresponding with the true states of affairs.

Kepler as well as Newton attempted to liberate themselves from a too narrow *logical* approach. Logically it is true that an abundance of hypotheses can be imagined to explain the phenomena. But neither Kepler nor Newton were interested in logically possible explanations. Kepler and Newton wanted to find the *laws*, given by God, determining the phenomena. The explanation of phenomena became secondary — the primary aim of science became for them the discovery of natural laws (see Sec. 12.2).

Galileo

The protests by Ramus and Kepler were by no means sufficient to refute Osiander's instrumentalism. Osiander happened to have a quite strong argument for his view. He observed that if Ptolemy's or Copernicus' construction of the motion of Venus were true, the brightness of Venus would vary much more than it actually does. The distance from Venus to the earth would vary by a factor of about 4, hence the apparent cross-section and the brightness varies by a factor of 16.[8]

This important argument was refuted only by Galileo's discovery of the phases of Venus in 1610.[9] Osiander's reasoning tacitly assumes Venus to be a primary source of light.[10] If, however, Venus merely reflects the light of the sun, it can easily be explained why the apparent brightness of Venus changes little, even if its apparent cross-section varies by a factor of 16, according to Osiander, or even 40, according to Galileo. When Venus is closest to the earth, it turns its dark side to the earth, whereas it shows its light side when it is far away.[11]

Galileo used his reasoning as an argument in favour of Copernicanism, although the same explanation can be given with Tycho Brahe's system. It refutes Osiander's instrumentalist argument, but cannot count as an argument in favour of Copernicanism.

Galileo rejected any instrumentalist interpretation of Coperni-canism (Secs. 3.3, 9.2). Such an interpretation was forced upon him by Cardinal Bellarmine, in 1616, when Galileo came into contact with the Inquisition for the first time. But Galileo preferred to be silent for a while rather than to embrace instrumentalism. When, in 1632, he again discussed Copernicanism, his choice was evi-dently in favour of a realistic interpretation of astronomy, not-withstanding his statement in the preface to the *Dialogue*: "...I have taken the Copernican side in the discourse, proceeding as with a pure mathematical hypothesis and striving by every arti-fice to represent it as superior to supposing the earth motionless — not, indeed, absolutely, but as against the arguments of some pro-fessed Peripatetics..."[12]

In Galileo's days, theologians discerned three attitudes to-wards hypotheses. These could be taken absolute, indeterminate, or hypothetical.[13] Concerning the hypothesis of the moving earth, the first attitude means the assertion that the earth moves. This was considered objectionable for lack of proof. The second attitude, meaning that there is neither proof that the earth moves nor that it is motionless, was practiced by Galileo in the *Dialogue*. Galileo did not prove the earth's movement, but instead he showed that the Aristotelian arguments against the earth's motion are invalid.[14] This attitude was also considered objectionable because the Bible teaches that the earth is im-movable. Only the third attitude, discussing the thesis "the earth moves" as a logical possibility, was acceptable. But the Co-pernicans wanted more.

Copernicus

In the *Dialogue*, Galileo made himself comfortable by omitting a great deal of astronomy. Galileo simplified the Copernican sys-tem to a set of concentric circles with the sun at its centre, and the Ptolemaic system to a set of deferents and epicycles. But both systems were far more complicated. In order to obtain a fairly ac-curate correspondence between calculated and observed planetary positions, both contained excenters, several epicycles for each planet, and Ptolemy also applied the principle of the equant.

As we have seen, Galileo compared his simplified version of the Copernican system with Aristotle's homocentric system, ra-ther than with Ptolemy's heterocentric one. This may be caused by the fact that his conservative colleagues at Padua were adherents to Averroës, the 12th-century Arabic commentator on Aristotle's

works. Being realists, Averroists tended to take Aristotle's cosmology to be literally true, rejecting the Ptolemaic heterocentric system in favour of Aristotle's homocentric one (see Sec. 2.1). Although the title of Galileo's *Dialogue* announces the comparison
of Copernicus' with Ptolemy's system, Galileo preferred to fight
the Aristotelian one. Unlike Ptolemy's, the Aristotelian system
was realistically interpreted, and Galileo was more interested in
cosmology than in astronomy.

Unlike Galileo, Copernicus wanted to make accurate calculations, and therefore could not afford to omit the details, except in
his popular expositions in the *Commentariolus* and the first book
of the *Revolutionibus*. He had to speak far more carefully about
hypotheses, because the excenters and epicycles in his calculations were instrumentally interpreted just like Ptolemy's devices.

Therefore, Copernicus could not object to Osiander's instrumentalist preface. He did not agree in principle, of course, but he
could not refute the sounder of Osiander's arguments. Galileo, Kepler, and Newton were able to do so — Galileo by refuting the
Venus argument, Kepler by making the excenters and epicycles
superfluous, and Newton by explaining Kepler's laws of planetary
motion.

11.2. The hypotheses of Copernicus and Tycho

In order to bring home the distinction between a hypothesis and an
axiom, we shall discuss a "rational reconstruction" of the way
Copernicus may have derived his theory from Ptolemy's. We call
this a "rational reconstruction" because it has little or no ground in
Copernicus' writings, and thus it is rather speculative, though,
hopefully, plausible.[15]

If we, like Galileo, waive many details shared by the two
systems, Ptolemy's can be reduced to two circles for each planet —
a deferent and an epicycle. The earth is at the centre of each deferent, and the planet travels along its epicycle, whose centre moves
along the deferent. By suitably choosing the periods, initial conditions and relative dimensions of the two circles, it is possible to
describe the observed phenomena — retrogradation and its coincidence with maximum brightness and opposition with respect to
the sun (see Sec. 2.1).

Copernicus' first modification of Ptolemy's system
In our reconstruction, we assume that initially Copernicus accepted
this geometric system. The astronomer is free to choose the radii

of the two circles, if only their ratio confirms to the observed size of the retrograde motion.

Now Copernicus makes use of this freedom, by choosing the *epicycle* of the *superior* planets, Mars, Jupiter, and Saturn, to be equal in size to the sun's orbit around the earth. Similarly, Copernicus makes the sun's orbit the *deferent* of the *inferior* planets, Venus and Mercury. From Ptolemy's calculations it was known that the periods in these five orbits are always exactly one year. Moreover, Ptolemy and his commentators knew that the radius vector between the centre of the epicycle of each superior planet and the planet itself is always parallel to the radius vector between the earth and the sun.[16] The inferior planets never move far away from the sun.

This "arbitrary" choice, made by Copernicus, may be considered a mathematical *hypothesis*, neither contrary to Ptolemy's theory, nor affecting the accuracy of the calculations. From this *ad-hoc* hypothesis, it follows that the inferior planets' epicycles have the sun as their centre. They move around the sun, and together with the sun they travel around the earth. For the superior planets, it is a little more complicated. Because the radius vector in the epicycle is always parallel to the line connecting the sun with the earth, as follows from observations, and is also equal to this line, as follows from the hypothesis, the distance from the sun to the planet is constant during the motion.

In other words, Copernicus' *ad-hoc* hypothesis implies that besides Mercury and Venus each other planet also moves in a circle with the sun at the centre. Nevertheless, the system is still Ptolemaic and geostatic.

Tycho's system

In order to arrive at Tycho's system (see Sec. 2.5) Copernicus' *ad-hoc* hypothesis must be replaced by another one, but only for the superior planets. Tycho interchanged the parts played by the deferent and the epicycle. The deferent is now the orbit of the sun around the earth, and the epicycle is a circle with the sun in its centre. Also this mathematical hypothesis is admitted in the Ptolemaic theory. This solution, at least with respect to Mercury and Venus, was already put forward by Heraclitus (4th century B.C.), and was known as the "Egyptian system."

There is no need to consider either one of the two *ad-hoc* hypotheses to be true. They are not even likely. They do not constitute anything more than an interesting logical or mathematical possibility to account for the coincidence of retrogradation, opposition and maximum brightness. They are merely *variants* of Ptolemy's system.

Copernicus' new theory

So far in our rational reconstruction, Copernicus and Tycho did not do much more than propose a variant of the Ptolemaic system. Tycho was content with it, for he shrank back from the idea of a moving earth. For quite a long time, many people considered the Tychonian system to be an acceptable alternative to the Copernican one, and it was even realistically interpreted (see Sec. 9.2).

But Copernicus was not first of all interested in a system for calculations, in a mathematical model able only to describe planetary motion. He wished to find a new *physical* system, a system allowing of giving explanations (see Sec. 2.4). He achieved this aim by designing a new *theory*.

This theory is new because it starts from a new axiom, namely: "All the spheres revolve about the sun as their mid-point, and therefore the sun is the centre of the universe."[17] Copernicus considered this axiom to be more important than the axiom concerning the daily motion of the earth. Other axioms are similar to Ptolemy's, including the axiom of uniform circular motion. It is only after the introduction of the *annual* motion of the earth that Copernicus had to introduce the *daily* rotation. The latter was considered many times before him (see Sec. 2.1), and was far less revolutionary than the hypothesis of the annual motion. It has no really important advantages over the geostatic theory. The apparent daily motion of the sun and the other planets can be explained either by the diurnal motion of the heavens, or by the diurnal motion of the earth, as long as the earth is assumed to be at the centre of the cosmos. Only in a heliostatic system does the explanation by the diurnal motion of the heavens become untenable.[18]

The new axiom differs from Copernicus' initial *ad-hoc* hypothesis because the earth is now considered a planet. It has also a different status — it is *true*. It is not true in an absolute sense, whether self-evident or provable, but in the sense that without the axiom the theory cannot stand. Copernicus' new theory is not characterized by uniform circular motion, but first of all by the new axiom, the axiom of the moving earth. (Therefore, Kepler was right to consider himself a true Copernican, though he refuted another axiom of Copernicus' theory, the axiom of uniform circular motion.)

Now we see clearly the difference between an axiom and a

hypothesis. An axiom is characteristic for a theory. If an axiom turns out to be false, the theory fails. An axiom is considered to be true, a hypothesis is a mere possibility. A hypothesis can be introduced as a boundary condition, in order to arrive at a better agreement of calculations and observations, but it does not determine the character of the theory. It may be true, but it does not need to be true. If a hypothesis turns out to be false, the theory need not be rejected, we only have to look for a new hypothesis in the context of the theory.

Nobody is committed to a hypothesis. But people may become committed to an axiom, as the Copernicans were committed to the axiom that the earth moves.

11.3. Descartes on hypotheses

Among the Copernicans, Descartes and Leibniz stand a bit apart. Both are considered to be philosophers rather than scientists, in the modern sense of these words. (During the 17th century, this distinction was never made, and the words "scientist" and "physicist" did not even exist. Until the end of the 18th century, every scientist was called a "philosopher.")

It may be doubted whether Descartes was a true Copernican. He wrote a book, *Le Monde* (The world), in which he assumed the motion of the earth. In a letter to Mersenne (1633), Descartes wrote: "If this (*i.e.*, the Copernican system) is false, all foundations of my philosophy would be false as well, for it (the motion of the earth) is evidently demonstrated from them."[19] But in the same year, learning of Galileo's conviction, he withdrew the manuscript from the printer. As a pious Roman-Catholic he did not want to challenge the church.[20] In later works, he presented himself as a Copernican in disguise, "...denying the motion of the earth with more care than Copernicus, and with more truth than Tycho Brahe."[21] According to his theory the earth moves around the sun because it is dragged by the whirling matter around the sun (see Sec. 3.4). But the earth is at rest with respect to this vortex, and therefore rests with respect to its direct surroundings. Calling this view "a hypothesis or supposition which is perhaps false", Descartes gave assurance that he respected the church's doctrines.

Light
The full title of Descartes' book reads: *Le Monde, où Traité de la Lumière* (The World, or Treatise on Light). In 1637 Descartes pub-

lished an introduction to his philosophy, the famous *Discours de la Méthode* (Discourse on Method). This introduction was accompanied by three treatises, serving as examples to show how his method should be applied. Besides mathematics in *La Géometrie*, and atmospheric phenomena in *Les Météores*, he studied optics in *La Dioptrique*. In the latter book he explained the refraction of light.

It is striking that nearly all leading Copernicans were concerned with optics, besides astronomy and mechanics. Kepler, Galileo, Beeckman, Descartes, Fermat, Huygens, Hooke, Boyle, and Newton, all contributed to the development of this field of science.[22] This, of course, is no accident. Until the discovery of radio-astronomy astronomy was observationally completely dependent on optics. The theories of optics always belonged to the presuppositions of astronomical theories. Even before his *Astronomia Nova* (1609), Kepler published *Astronomia Pars Optica* (The optical part of astronomy, 1604), one of his two major optical works.[23] Also Newton's early investigations of light were related to astronomy. The question whether lenses could be made without colouring defects led him to his experiments with prisms, and the conviction that the problem was unsolvable. Therefore he designed the reflecting telescope, stressing that it lacked the defects of the refracting telescope. A model of his invention sent to the Royal Society (1672) made him famous at one stroke. This induced him to publish his experiments on prisms (see below).

But there may have been another reason why the Copernicans particularly showed interest in the theory of light. All of them rejected the fundamental Aristotelian distinction between heavenly and earthly phenomena. According to Aristotle, the four elements are strictly confined to the sublunar spheres, and the celestial spheres consist of some other material, *aether* or *quintessence*, the fifth element (Sec. 3.1,3.2).

But if this distinction is so radical, how are we able to *see* a celestial body? Is light a kind of matter, existing both in superlunary and sublunary spheres? Or is aether, the bearer of light, present in both regions? The problem of light has always been an embarrassment in Aristotelian physics.[24] This may be the second reason why Copernicans were eager to study optics. Here they thought to catch the Achilles' heel of the detested Aristotelian philosophy.

For Descartes, another reason to pay much attention to light was his view that "seeing" is the most important way of perceiving.[25] Hence the relation between object and observing subject is mediated by light. Descartes not only contributed to the physics of seeing, but also to its physiology and psychology.

Descartes' hypotheses

Because light and seeing play a central part in Descartes' philosophy, his theory of light should start from a "clear and distinct idea." It is that light is propagated instantly, with an infinite speed, through a medium pervading all other kinds of matter. Because Descartes identified matter with space, he could not conceive a void. The only kind of motion in a plenum is vortical motion (Sec. 3.4), and light propagating rectilinearly cannot be motion. Therefore, it is inconceivable that light would need time to move from one place to another, and visual perception has an immediate character. In 1634, Descartes wrote to Beeckman: "To my mind, it (*i.e.*, the instantaneous motion of light) is so certain that if, by some impossibility, it were found guilty of being erroneous, I should be ready to acknowledge to you immediately that I know nothing in philosophy...if this lapse of time could be observed, then my whole philosophy would be completely upset."[26]

It is therefore quite remarkable that in his *Dioptrique* Descartes assumes that light does have a finite speed, different in various media. With this assumption he is able to explain refraction, and to derive Snel's law.[27]

This is one of the most disputed features of Cartesian physics. Descartes has some axioms which he puts forward on metaphysical grounds, and in which he believed without any reserve, because they are "clear and distinct." But he is aware that these principles are much too simple to explain the full complicated reality. If he wants to give an explanation of some phenomenon as plain as refraction, he concocts a hypothesis — in this case the hypothesis that light in various media moves at different speeds — with the sole purpose of demonstrating that the new mechanical philosophy is able to explain everything.

In fact, however, Descartes does not assert that light actually moves. He says that light has a *tendency* to move. Snel's law is derived by making an *analogy* with a really moving *ball*, having various speeds in different media.[28] He only wants to suggest a *possible* explanation.

Descartes considered it of no relevance that his hypotheses contradicted his own clear and distinct axioms, and therefore were false.[29] He maintained the medieval doctrine of double truth mentioned in Sec. 2.1. It is the task of metaphysics to give explanations which are true, and the task of physics to find theories derived from false hypotheses but describing the phenomena correctly. The only difference from Aristotle's adherents is the terminology. Because Descartes refused to distinguish between physics and mathematics, or physics and astronomy, he had to divide the tasks between physics and metaphysics.

The physicists (in the modern sense) soon took his advice to heart. They accepted the separation of physics and metaphysics, leaving metaphysics to the philosophers. But they did so in a way quite different from Descartes' intentions. The physicists started to consider the content of their physics to be true, and they left Cartesian metaphysics for what it was — rationalistic speculation. Soon they arrived at the conclusion that Descartes' clear and distinct ideas were untenable, and that their own physics supplied more certainty than any metaphysics. Physical questions should be settled by physical means.

The first fruit of this attitude concerns the speed of light. Already Galileo had doubted that the propagation of light would be instantaneous. Whether this was correct or not could only be found by an experiment, but his initial trials to determine the speed of light failed.[30] He could not know that his own discovery of Jupiter's moons would enable Rømer and Huygens to show that the speed of light is finite, and to determine its value (1676).

11.4. Hypotheses non fingo

Newton's *Opticks* (1704) begins with the words: "My design in this book is not to explain the properties of light by hypotheses, but to propose and prove them by reason and experiments."[31] It would be impossible to put it more clearly. Newton rejects the use of hypotheses.

In fact, however, this and similar statements are far from clear, and it is a much disputed matter what Newton exactly understood by hypotheses.[32] Before 1687, when he published his *Principia*, Newton did not bother very much about this philosophical problem, and in the first edition of the *Principia* he used the word "hypothesis" ambivalently. He only started to reflect on the status of hypotheses under the influence of criticism directed at the *Principia*.

Opticks
Strictly speaking, Newton had come into conflict about the use of hypotheses much earlier. In 1672 he published his experiments about the refraction of light by a prism.[33] He arrived at the con-

clusion that white light is a heterogeneous mixture of "coloured" rays.[34] Colour does not arise from an interaction of white light with matter, but is inherent in light. Assuming that this conclusion follows exclusively and irrefutably from his experiments, Newton thought that no hypothesis about the nature of light underlies it.

His paper was met with enthusiasm, and his experiments were highly praised. But Hooke and Huygens maintained that Newton did make use of hypotheses, and that their own alternatives could explain the experiments without accepting the heterogeneity of white light. Newton rejected this view, and in 1675 decided not to publish anything else. Fortunately, he did not hold to this resolution.

Queries

Whereas in 1687 Newton was still rather uncritical in his use of the word "hypothesis", in the *Opticks* (1704) and in the second (1713) and third (1726) edition of the *Principia*, he was far more careful. Where in the first edition of *Principia* Newton used the word "hypothesis", in the later editions he nearly everywhere replaced that by "rule", or "phenomenon", or "theorem."

The cases where he maintained the word "hypothesis" are for statements which he assumed to be unproved or uncertain. These concern, first, the internal friction in a fluid; second, the so-called "Copernican hypothesis", *i.e.*, Kepler's first and second laws; next, the hypothesis that the centre of the world is immovable; and finally, a hypothesis concerning the motion of a material ring around the sun.[35] Calling Kepler's laws hypotheses is nearly as curious as Newton's statement at the end of the book, immediately after his determined statement "I frame no hypotheses", concerning "a certain most subtle spirit which pervades and lies hid in all gross bodies," which he does not call a hypothesis.[36] "But", he adds, "these are things that cannot be explained in few words, nor are we furnished with that sufficiency of experiments which is required to an accurate determination and demonstration of the laws which this electric and elastic spirit operates."[37]

The latter words were written in 1726, in the third edition. Earlier, in the *Opticks*, Newton had already speculated about the "aether." In this book he also avoided hypotheses, except at the end, where he included a number of problems, or questions: *Queries*. In these Queries quite a few hypotheses occur, bearing witness to Newton's great imagination. The first English edition

(1704) contains sixteen Queries. The Latin translation (1706) added seven more, and the second English edition (1717-18) contains thirty-one Queries in all. Though mostly dealing with light, the Queries also concern other subjects. During the 18th century, English Newtonians in particular were strongly inspired by these Queries, which Newton proposed"...in order to a farther search to be made by others."

The final and longest Query[39] contains a statement of method: "As in Mathematicks, so in Natural Philosophy, the Investigation of difficult Things by the Method of Analysis, ought ever to precede the Method of Composition. This Analysis consists in making Experiments and Observations, and in drawing general Conclusions from them by Induction, and admitting of no Objections against the Conclusions, but such as are taken from Experiments, or other certain Truths. For Hypotheses are not to be regarded in experimental Philosophy. And although the arguing from Experiments and Observations by Induction be no Demonstration of general Conclusions; yet it is the best way of arguing which the Nature of Things admits of, and may be looked upon as so much the stronger, by how much the Induction is more general. And if no Exception occur from Phaenomena, the Conclusion may be pronounced generally. But if at any time afterwards any Exception shall occur from Experiments, it may then begin to be pronounced with such Exceptions as occur. By this way of Analysis we may proceed from Compounds to Ingredients, and from Motions to the Forces producing them; and in general, from Effects to their Causes, and from particular Causes to more general ones, till the Argument end in the most general. This is the Method of Analysis: And the Synthesis consists in assuming the Causes discover'd, and establish'd as Principles, and by them explaining the Phaenomena proceeding from them, and proving the Explanations."[40]

Hence, whereas he stresses that induction is not proof, Newton states that one should hold to its results as long as no exceptions are found.

Hypotheses non fingo

The famous statement "I frame no hypotheses", or "I feign no hypotheses" cannot be found in the first and second editions of *Principia*, but only in its third edition (1726), and thus in its translation into English by Motte (1729). At the end of the book, in the General Scholium, Newton writes: "But hitherto I have not been able to discover the cause of those properties of gravity from phenomena, and I frame no hypotheses; for whatever is not deduced from the phenomena is to be called a hypothesis; and hypotheses, whether metaphysical or physical, whether of occult qualities or mechanical, have no place in experimental philosophy. In this philosophy particular propositions are inferred from the phenomena, and afterwards rendered general by induction. Thus it was that the impenetrability, the mobility, and the impulsive force of bodies, and the laws of motion and of gravitation, were discovered. And to us it is enough that gravity does really exist, and act according to the laws which we have explained, and abundantly serves to account for all the motions of the celestial bodies, and of our sea."[41]

The cause of gravity

For Aristotle, gravity was the natural tendency of heavy bodies to move to their natural place, the centre of the universe.[42] Galileo professed not to know what gravity is, except that it is an inherent property of bodies, and the source of motion.[43] But he found a quantitative law concerning the fall of bodies in a void. Descartes explained gravity as caused by vortical motion (Sec. 3.4).

According to Newton, gravity is not an inherent property of a body, but a *relation* between bodies. It is like a state of motion, and unlike mass, which is an inherent property, quantity of matter, inertia. Gravity is both independent of and irreducible to motion. Like Galileo, Newton states that he does not know what the essence of gravity is, and he does not frame hypotheses.

For this reason, Newton is sometimes said to be a positivist *avant la lettre*, but this is surely mistaken. For, though Newton admits he does not know the cause of gravity, he knows its laws, and "...to us it is enough that gravity does really exist..." This is a realistic view.

Because he introduced action at a distance, Newton was charged to take recourse to "occult" principles of explanation. Newton wards off this blow by equally rejecting "occult" and "mechanical" qualities, if they are not based on carefully analyzed phenomena — and everybody understood "mechanical" to refer to Descartes' explanation of gravity.

What is against hypotheses?

Philosophers like Hempel and Popper, swearing on the hypothetical-deductive method, and also having a high opinion of Newton, may feel embarrassed by Newton's firm rejection of hypotheses, and his insisting on "...propositions (which) are inferred from the phenomena, and afterwards rendered general by induction."

Still, Newton was not an inductivist, nor did he reject all kinds of hypotheses. He did not object to hypotheses which can be connected with phenomena. He considered the proposal of a hypothesis fruitful if it leads to further research, to new observations and experiments, and to the development of theories.

Newton rejected the assumption that axioms could pretend to contain evident truth. Newton opposed Cartesian metaphysics with its infallibly clear and distinct ideas. He stated that concepts like hardness, impenetrability, and mobility are found on the basis of phenomena, and are by no means evident properties of matter.[44] Newton did not care that his law of gravity is not clear and distinct; "...it is enough that gravity does really exist, and act according to the laws which we have explained, and abundantly serves to account for all the motions of the celestial bodies, and of our sea."

In particular, Newton rejected the use of hypotheses introduced in the framework of a double truth practice. Newton did not adhere to the instrumentalist view that statements in a theory need not be true if only the phenomena are described correctly. Newton never accepted that his law of gravity could be false, unless it proved to be unable to explain the phenomena. He recognized that his theory is limited in scope, but he held to its correctness, unless it would clash with observation and experiment. For the latter constitute the decisive evidence, rather than an authoritative theology, or a supposed self-evident insight.

Finally, Newton rejected hypotheses as having a definitive character. Hypotheses may be useful to suggest new experiments at an early stage of research. But they can never be acceptable in a more or less definitive theory, to which its adherents should be committed. A mature theory should start from axioms which are inductively found, and should have proved its fertility, its problem solving and problem generating capacity.[45]

Hence, according to Newton we should not look for unfounded hypotheses, but for axioms, which are firmly grounded in objective experience, and which are believed to be true. One should be committed to one's axioms.

Reliability

Reading Newton fulminating against those who frame hypotheses, one cannot escape the feeling that he was very angry with them. Possibly he considered them unreliable.

Science is a communal endeavour, and can only proceed if one scientist may rely on the results of his fellow scientists. Studying Descartes' *Principles of Philosophy*, Newton found his physical theories highly unreliable. As we have seen, he devoted the second book of the *Principia* to refuting these theories, which he not only considered to be wrong, but below the standard he was to establish himself.

Commitment and reliability are related to each other, for who would trust a scientist who does not stand behind his own theories? How could one accept a theory, if its inventor cannot be relied to have investigated its evidence, to have considered its most obvious consequences, to be convinced of its consistency, to believe its basic axioms?

Therefore, commitment to a theory is not something to be tolerated, as an understandable human weakness — it is a necessary ethical norm, to which every scientist should conform if he wants to be taken seriously, to be trusted by his fellow scientists.

However, we should refrain from absolutizing any aspect of human experience, including the ethical aspect. No scientist is required to hold fast to his or her theories at any cost. In particular, scientists have to take into account the criticisms leveled by other scientists. The ethical norm of commitment means that any scientist should take his or her responsibility seriously, and this includes the eventual reconsidering of one's views.

12. Belief

12.1. The search for certainty

The final chapter of this book is concerned with the question of what certainty the use of theories may provide. This question is related to the problem of the meaning of concepts and the truth of statements. Various philosophies have tried to reduce the problem of certainty to meaning, or to truth, or to consistency, or to usefulness. We shall attempt to make clear that it makes little difference, because the meaning of concepts, the truth and consistency of statements, and the usefulness of theories are mutually dependent.

The first section is mainly concerned with past views on universals. In the middle ages, three views can be distinguished, idealism, realism, and nominalism or voluntarism, or instrumentalism.[1]

Idealism

In the 13th century, idealists (mostly Franciscans) were impressed by Plato's doctrine of ideas[2] (see Sec. 3.2). They considered the nature or essence of things to be determined by abstract *ideas* which transcend the concrete things. The ideas logically precede concrete being, they are *"ante rem"*, before being.

Although by any standard Plato was not a mathematician, his views on ideas were mathematically oriented. He was more or less a disciple of Pythagoras (see Sec. 3.1). In order to understand medieval idealism we must think of geometry. In our daily life we encounter triangular objects, which, strictly speaking, are never truly triangular. As defined and studied by geometricians, the real, ideal triangle only exists in abstract thought. As a consequence, the abstract ideas determining the essence of things can only be understood in thought. Observation has little impact on idealistic thought. Sensory experience is deceptive.

Especially during the first half of the Copernican revolution

Platonism exerted a strong influence, for instance on Copernicus, Kepler, Benedetti, Galileo, and Descartes. This influence was transmitted via Archimedes, another Platonist, whose works were published in 1543, almost simultaneously with Copernicus' *Revolutionibus*. Archimedes made a strong impression because of his mathematical treatment of physical problems, and because of his thought-experiments.

In Platonism, theoretical statements express abstract ideas. Hence, Copernicans who were influenced by Platonic ideas discussed idealized problems: the rigid lever, frictionless motion, and the fall in a vacuum, for example.

Galileo's Platonism is a disputed matter,[3] and of course he denied many Platonic teachings. But it should not be overlooked that he deliberately applied Plato's literary form of dialogues, both in his *Dialogue* (1632) and his *Discorsi* (1638).[4] He also distrusts appearances: "It is therefore better to put aside the appearance, on which we all agree, and to use the power of reason either to conform its reality, or to reveal its fallacy." Galileo gives an example "...from which one may learn how easily one may be deceived by simple appearances, or, let us say by the impressions of one's senses."[5] Galileo admits that human reasoning is less perfect than Divine intuition, but only as to its extension. Nevertheless, "...human intellect...has as much absolute certainty as Nature itself has. Of such are the mathematical sciences alone; that is, geometry and arithmetic...with regard to those few which the human intellect does understand, I believe that its knowledge equals the Divine in objective certainty."[6]

But, as observed in Sec. 4.5, Platonism was mainly used as an ally in the fight against Aristotelianism.

Realism

Disciples of Aristotle, medieval realists can be found both among Arab scholars like Averroës and among the Dominican friars like Thomas Aquinas. The nature or essence of things is first of all determined by their "forms" (see Sec. 3.2), and in this respect there is not much difference from Plato's ideas. But Aristotle held that form and matter, both eternal and universal, constitute a unity in each individual thing or "substance." Hence we can obtain knowledge about "universals" by studying individuals. The forms are present in being, they are *"in re."* For realists, observation is an important source of knowledge (see Sec. 5.3).

For Plato, change is irrational, it belongs to the realm of deceptive appearances. The only real things, the ideas, are unchangeable. Aristotle, however, relates "nature" both with "forms" and with "change." His *Physics* is concerned with the theory of change (see Sec. 3.2): "Nature is the distinctive form or quality of such things as have within themselves a principle of motion, such form or characteristic property not being separable from the things themselves, save conceptually",[7] but also "Nature is the principle of movement and change."[8]

Unlike Plato, Aristotle was not inspired by geometry but by biology. He was an expert biologist, and did pioneer work in the classification of the animal kingdom into species and genera. He was aware of the fact that the "form" of animals, *i.e.*, the properties determining their essence, their species, cannot be found by thought alone. Nevertheless, Aristotelian realism, too, is rather rationalistic. In principle all forms are rational, and observation was only called upon to help if human thought fell short.

The Aristotelians became very influential especially after Thomas Aquinas' accommodation of Aristotelian philosophy to the doctrines of the church. From the 13th century Thomism was to remain the official philosophy of the Roman Catholic church until the end of the second world war. In the 16th century, during the rise of Copernicanism, most universities were dominated by Aristotelian philosophers (see Secs. 8.1, 8.2).

The realist view of "truth" is expressed in the Latin words *"adaequatio rei et intellectum"*, the correspondence of being and thought. The meaning of terms was to express the essence of things. This view is shared by idealists, by Galileo ("No greater truth may or should be sought in a theory than that it corresponds with all the particular appearances"[9]), and perhaps surprisingly, by Popper.[10]

Instrumentalism

Whereas the 13th-century Franciscan friars were often idealists, during the 14th century many Franciscans became nominalists. They opposed the rationalism inherent in the doctrine of ideas or forms. Both Plato and Aristotle assumed that ideas or forms cannot be other than they are, because they are rational. William of Ockham and other nominalists thought this to be contrary to the Christian doctrine of God's sovereignty. Their views, even upheld by Pope Urban VIII against Galileo (see Sec. 9.2), originate probably with the Islamic view of God's omnipotence. It is called "voluntarism", and says that God's will is unpredictable, absolutely unrestricted. (*"Islam"* means "surrender", namely, to the will of God.)

For instance, Plato and Aristotle taught that celestial bodies can move uniformly only in circular orbits around the centre of the universe. The nominalists suggested instead that God could have made the world differently if he had wished to do so; thus celestial bodies could have moved otherwise.

Extreme nominalists rejected the reality of universals entirely except in human thought. They said that universals are invented by people to facilitate thought and observation. Outside thought only individual concrete things and events can be found, which therefore precede the universals. The universals are logically *"post rem"*, after the things. Hence, for the nominalists observation is even more important than for the realists. We observe similarities which we generalize with the help of universals. For instance, we observe so many similarities between Venus, Mercury, Mars, Saturn, and Jupiter, that we find it easy to subsume them under the same universal name, "planet." But independent of the concrete planets, there is nothing like an "idea" planet, or a "form" planet. "Planet" is merely a name (Latin: *nomen*), just like "Mars" is only a name.

During the 14th century, in particular at Paris, several nominalists developed theories about motion anticipating those of Galileo (see Sec. 3.1). For this reason it has been assumed that nominalistic influences were strong during the Copernican revolution. If we include Plato among the realists, the Copernicans were realists rather than nominalists, however. A nominalist view on theories is called *instrumentalism*. It says that theories are nothing but instruments useful for the prediction of phenomena, for instance, the observed celestial motions. Instrumentalists have certainly played a part in the Copernican revolution, as we have seen, but always in the camp of the opposition. Copernicans nearly always rejected instrumentalism.

A modern version of instrumentalism is Mach's philosophy of science (Sec. 9.1), and Laudan's view that problem solving is the sole aim of science.[11] Laudan denies that science searches for truth. But he overlooks the widespread scientific belief that the solution of a problem is only satisfactory if it is considered to be a *true* solution, whatever the way by which its truth is established.

The probability of concepts and the concept of probability

Aristotelian and Cartesian rationalistic philosophies were challenged by Pascal, Boyle, Newton and others. During these discussions, the modern concept of probability was born.[12]

Medieval Aristotelians distinguished between statements

whose truth is certain and other statements which are at most probably true. *Certain* statements were either based on the Bible, the theology of the church fathers, the doctrines of the church, or on direct self-evident principles. Such self-evident truths were often taken from Aristotle or other authorities. An example is the insight that a void cannot exist, for what is not cannot be; or that the celestial bodies are perfect and incorruptible; or that every change needs a cause. Some Cartesian "clear and distinct" — hence evidently true — statements are the proposition that matter is identical to space and therefore that a void cannot exist, and the proposition that light is propagated instantaneously and thus with infinite speed.

In the 17th century many people began to doubt the possibility of obtaining absolutely true insights on the elementary ground that many if not all of these self-evident truths turned out in the end to be false. It was not on philosophical, but on experimental grounds that Pascal, Boyle and Newton argued that a void exists. Newton, for instance, pointed to the fact that the planets move without any friction around the sun, which he thought to be inexplicable without the assumption that interplanetary space is void. Tycho and Galileo showed that celestial bodies are no more perfect and incorruptible than the earth. Rømer demonstrated that the speed of light is finite, and measurable.

Statements not covered by one of the above mentioned authorities were considered *"probable"* (Sec. 11.1). Whereas certain truths and their consequences were studied in *science* (*e.g.*, physics), probable truths belonged to the *liberal arts* (*e.g.*, astronomy). Attempts were made to derive criteria in order to accord a degree of probability to statements. The attention drawn to this view of conceptual probability may have retarded the development of a modern view of probability. Only because most Copernicans rejected the distinction between "evidently true" and "probable" *statements* could they pay attention to the probability of possible equivalent *events*, such as occur in games of chance, or such as are covered by insurance. In the 17th century, Fermat, Pascal, Huygens, and others were interested in games of chance, and many municipal authorities sought their advice in problems concerning life insurance. This development was also furthered by the shift from the Aristotelian emphasis on qualitative distinctions to the Copernican search for measurable quantities.

In the 20th century, positivist philosophers have tried to find a measure of the probability of statements.[13] They rejected the possibility of finding certain knowledge concerning empirical states of affairs, but they tried to anchor the reliability of scientific statements in observations. It amounts to the idea that a hypothesis is more probable if it is confirmed by more observations. The elaboration of this idea has caused many problems, and it may be concluded that it is impossible to find a measure of the probability of statements.

We have seen that Popper radically rejects this positivist view (see Sec. 2.3).[14] According to him, science must not look for probable but for highly unlikely hypotheses. It should not be attempted to confirm them, but to refute them, not to verify but to falsify one's conjectures. This criticism did not restrain Popper from trying to derive a formula to calculate the "degree of corroboration" of any statement. But he adds that he does not believe that his formula makes a positive contribution to science, methodology, or philosophy.[15] We cannot but agree.

Meaning, truth, and the law of non-contradiction

There is some confusion about the mutual relation of meaning, truth, consistency, *etc.* (see Sec. 1.5). In a *logical* sense, it is not really difficult, however, to demonstrate their distinction and relation. We do this by referring to our distinguishing between concepts, statements, and theories. We find then that the norm of "meaning" applies to *concepts*. The medieval discussions, summarized above, were mainly concerned with the meaning of universals, in particular class concepts. Idealism, realism, and nominalism differ because of the status of "meaning" they attribute to the universals.

The norm of "truth" is specifically related to *statements*. A statement may be true or false, and this partly depends on the meaning of the concepts contained. Hence, the distinction among idealism, realism, and nominalism, originally directed to universals can be transferred to statements, in particular law statements. In Chapter 11 we have discussed the different status of hypotheses, whose truth is not asserted, and of axioms, whose truth is defended by those scientists who feel committed to them.

The most important norm of logic is the law of non-contradiction. It cannot be applied to a single statement (a statement cannot contradict itself, if we leave aside some typical paradoxes), but it concerns two or more statements, which occur in the same context. The most organized set of statements is a *theory*, a deductively connected set of statements. Hence, the most important norm for theories is the law of non-contradiction. But, as we argued in Sec. 1.3, this law can only function if the statements in the theory are supposed to be true within the context of the theory, and again, this assumes that the meaning of the concepts used is clear.

Hence, it is something like a mistake of categories to ask whether a theory is true. It could either mean that all statements in the theory are true in an absolute sense, which is never the case, or it could mean that its basic axioms are true. In both cases, the "truth" of the theory boils down to the truth of its elements. The relevant *logical* question is whether theories are free from contradictions, both internally and externally.

But though logically qualified, theories are applied predominantly for non-logical purposes. A theory should be able to predict and to explain, to generate and solve problems, to systematize our knowledge, and (as we shall see) to help us to develop the laws of nature. Besides we have seen that theories should be clear, useful, economic, and beautiful, and have juridical and ethical aspects. There is no reason to single out one of these aspects to be the most important. Therefore, there is not a single criterion for accepting or rejecting a theory. Theoretical work is not that simple.

12.2. The status of natural laws

In the present section we are concerned with the *ontological* status of laws. Positivists reject the distinction between laws and law statements, and thus the existence of laws outside human experience, for they assume that laws are no more than human inventions made to order our sensory experiences. We shall pay attention in particular to the reformational view that laws are given by God as the Sovereign of the creation, and are open to human investigation and exploration.

The search for order
The idea of "laws of nature" emerged during the Copernican revolution after 1600, probably without much deliberation. Kepler's laws belong to the oldest statements called "laws" in the natural sciences. Galileo wrote "...Nature...is inexorable and immutable; she never transgresses the laws imposed upon her, or cares a whit whether her abstruse reasons and methods of operation are understandable to men."[16] Descartes said that laws of nature are established by God, and he defined them as "Rules according to which changes take place."[17] And Newton introduced his basic axioms as "laws of motion" and "law of gravitation."

The increasing emphasis on "laws" cannot be understood apart from the general historical context. During the middle ages, the concept of law as we know it hardly existed. Countries were partly ruled according to agreements, contracts, between the rulers and the representatives of the people. For another part, the emperors, kings, *etc.* derived their authority from God, or from the church. After the Reformation, people started to consider laws to be based on fundamental ideas like justice, freedom, or human rights, and which therefore transcended both agreements and royal authority. A law transcends the authority of the government, even if the latter is the primary source of *positive* law, *i.e.*, the law as it is formulated in law books or constitutions. It means that the government is subject to its own laws. In 1581 the Dutch States General abjured their lord (the Spanish king) because he violated the country's laws.

The medieval practice of government by agreement collapsed under the burden of its complications and arbitrariness, its lack of unity. The idea of law arose during the religious and civil wars because of the generally felt need of order.

We find this need also in the physical sciences. Copernicus' main motive to reform astronomy was his wish to bring order in the planetary system. He criticized his precursors, saying: "Also they have not been able to discover or deduce from them the chief thing, that is the form of the universe, and the clear symmetry of its parts. They are just like someone including in a picture hands, feet, head, and other limbs from different places, well painted indeed, but not modelled from the same body, and not in the least matching each other, so that a monster would be produced from them rather than a man."[18]

The reformed view of law

The reformers Luther and Calvin shared the nominalists' criticism of idealistic or realistic rationalism (see above). But they feared that a one-sided emphasis on God's omnipotence would lead to the idea that God acts arbitrarily. In particular Calvin supplemented the idea of God's omnipotence with the idea of God's faithfulness. God is faithful to his covenant with his people, to the laws which he accorded the creation, including the natural laws.

The reformers rejected the Platonic and Aristotelian position which maintained that "ideas" or "forms" are logically transparent, self-evident, purely rational. This view was still shared by Galileo and Descartes. But Kepler arrived at the view that natural laws are neither logical nor intuitively evident. The planets move in elliptical orbits with a velocity changing according to a law — but this could have been different. Newton proved Kepler's laws from his law of gravity, but he could not logically prove this law to be necessary. The law of gravity is as it is, but it could have been different if God had wanted it so; therefore it is "contingent" on God's will. This is a nominalist view.

This view is now balanced by the reformational belief that God maintains his laws because he is faithful to his creation. This allows us to *discover* the laws. They are not first of all open to rational thought, but to empirical investigation, in which rational thought operates together with observation and experiment. Hence, in this view observation and experiment are even more important than in realism and nominalism. In nominalism, observation is the only source of knowledge, but any generalization is unwarranted, except as a practical means to the economic ordering of our experience. Kepler, Pascal, Boyle and Newton sought this warrant in God's promise to maintain the creation.

Hence, Kepler and Newton believed their law statements to be true. This belief was not based on observations. They knew very well that observations and experiments are fallible. Nor was it based on their reasoning, which they knew to be limited as well. But they believed that God has given unchangeable laws for the creation, and they believed that these laws are open to human investigation.

Modern realism
Whereas medieval realists emphasized the reality of "universals", *i.e.*, concepts or ideas, modern realists emphasize the objective existence of laws as a metaphysical principle. But their idea of "truth" is still the medieval idea of correspondence between theories (or hypotheses) and facts.

Thus, Bunge asserts "...the hypothesis that scientific research presupposes the metaphysical hypothesis (that objective laws exist). The reliability of objective patterns will be granted by anyone agreeing that the central goal of science is the *discovery* of objective patterns...", adding: "...law formulas are invented, but laws are discovered."[19] Popper also is a realist in this sense.[20]

This kind of realism which accepts the metaphysical principle of the objective existence of laws should not be identified with the religious view of the reformers and many 17th-century scientists, who not only accepted the real existence of laws, but also recognized the origin of these laws, and their firm foundation. This is not a matter of metaphysics, it is not a hypothesis to be tested, but a matter of belief.

The reformed view of law implies a different view on "truth." Truth is now not conformity of theory and fact, but law-conformity, obedience. The investigation of the lawfulness of the creation is conducted by respect for the laws, or rather for the lawgiver, the Sovereign of heaven and earth. It was this respectful attitude which led Kepler to accept his laws, which contradicted every hypothesis conceived up to 1600. It means the subordination of human thought to divine law.

The transcendental character of laws

Two historically important views of laws have failed. The first, represented by Descartes and other rationalists, held that laws must be reducible to clear and distinct ideas, in order to achieve a rational and necessary character. It failed ever since Newton recognized the law of gravity to have a contingent character. His rejection of rationalism has been reinforced by 19th- and 20th-century developments in the natural sciences. This means that natural laws transcend rational thought.

The other view is that of the inductivists, who assume that laws are nothing but generalizations of observations. Because natural laws are supposed to be valid everywhere and always, this inductivist view is untenable. This means that natural laws transcend human experience.

Law statements are invented by men, and spring from their imagination in a process including both rationality and experience. But the laws they refer to even transcend this imagination. Usually the far-reaching consequences of newly discovered laws cannot be foreseen, and are much richer than anybody could have predicted.

This threefold transcendental character of laws is radically different from Platonist transcendental idealism. In the latter case, observable things are imperfect copies of the real and perfect ideas. In Aristotelian realism, an observable thing is a unity of form and matter. It is imperfect as long as it has not actualized all its potentialities. It is notable that Descartes also relates the laws to perfect, clear ideas. He contrasts God as a perfect being with man, who is imperfect because of his doubt. But in the modern view of law, the idea of perfectness hardly plays a part. The observable things are not copies of laws, but are subject to laws.

Scientists still speak of "ideal" things, either in a conceptual sense (a rigid lever, an ideal or even perfect gas), or in an experimental sense (a purified sample, a thermostat). The meaning of these objects of research is not to obtain a more *perfect* sample, but rather a sample which is simpler than anything we find in nature. It is easier to do calculations on a pure ideal gas than on an impure mixture of oxygen and nitrogen. It is easier to do experiments in an enclosed room kept at a constant temperature than in an uncontrollable open space. The "idealization" used in present-day science is intended to eliminate disturbing circumstances.

12.3. World views

Whereas religion has its theology, science has an ideology. The presuppositions of a theory include a world view, a metaphysics. It is seldom explicitly put into words. If scientists elaborate on their world view, they do not do so in their scientific works, but in private, in semi-popular talks, or in their autobiographies. A serious study of a scientific world view is left to philosophers or historians of science. Nevertheless, it is of great importance.

Laudan prefers "research tradition" above "world view" or "paradigm." A research tradition does not have an explanatory, predictive or problem solving function, and cannot be tested. Its function is to determine problems, to indicate what kind of test is acceptable; it has a heuristic function. A research tradition identifies three kinds of suppositions: unproblematic, because accepted; forbidden; and suppositions which have to be explicitly justified.[21]

We shall consider four aspects of a theoretical vision of reality, the ontological (what does the world look like?), the epistemological (what are the sources of our knowledge?), the logical one (what counts as proof?), and the heuristic aspect (how are theories found?). Inevitably, we shall mostly be concerned with the mechanistic world view, or "mechanism" for short, another fruit of Copernicanism.

Mechanism (The mechanistic world view)

By no means should Copernicanism be identified with mechanism. Although Copernicanism gave rise to the insight that motion is an irreducible mode of experience, it differs from mechanism, which we shall define as the world view absolutizing this mode. Its prime, though not its first protagonist, is Descartes. Galileo, Beeckman, Huygens, and Leibniz may also be considered mechanists, but this is far less the case with Kepler and Newton, who, as we saw in Secs. 3.5 and 3.6, relativized mechanism by introducing another irreducible principle of explanation, physical interaction. The opinion that Newton was not a mechanist is not shared by all historians of science. In particular Dijksterhuis, though admitting that Descartes was the main representative of mechanism, and even pointing to Huygens as the most pure mechanist,[22] nevertheless considers Newton's work the acme of the "mechanization of the world picture"[23] (see Sec. 3.6). We shall not deny that Newton occasionally paid lip service to the mechanical philosophy, by which he was evidently influenced. But his criticism of Cartesian mechanism prevails.

The mechanistic world view only slowly replaced the ancient and medieval world view, aptly called "organicism." This world view was characterized by everything having its own natural place. The Aristotelian philosophy of nature fitted the medieval society with its double hierarchies of the Holy Roman Empire, and the Holy Roman Catholic Church very well. This hierarchical world view required the central (*i.e.*, low) position of the earth, and could therefore not be maintained by the Copernicans. Although alchemy and astrology are not of Greek origin, both could easily be accommodated in the organistic world view, their main principle being the correspondence between a macroworld and a microworld. Mechanism rejected both astrology and alchemy, and it is significant that among the Copernicans, only Tycho Brahe and Kepler accepted astrology, and Newton was an expert in alchemy. However, Kepler was critical of both astrology and alchemy,[24] and Newton did his chemical experiments in a reproducible way, carefully determining the quantities of his reagents.[25]

The world *picture* of mechanism is, of course, a machine. The clockwork model of the universe was very popular.[26] Yet we should take care to distinguish this *picture* from the world *view* of mechanism, if only because the clockwork model dates from the middle ages. One of the characteristics of the Aristotelian form-matter scheme of scientific explanation is its emphasis on *purpose*, on the "final cause", the *end*, as a principle of explanation. Now it cannot be denied that machines, too, are things which can be fully explained only if their purpose is taken into account. And, of course, speculation about the purpose of the universe as a machine is found especially in popular expositions of the mechanical philosophy. But in their scientific works, mechanists distanced science from the principle of a final cause. In particular Descartes rejected the search for final causes.[27]

The shift from organicism to mechanism implied a revaluation of "mechanics." In the middle ages and earlier, mechanics was a *craft*, not a science. It was developed to a high level of sophistication quite apart from academic philosophy. During the Copernican revolution, mechanics became the paradigm of science, the basis of any scientific explanation. Organicism saw mechanics to be concerned with artificial things like levers, pumps, and mills, unfit to explain natural states of affairs, which was the concern of "physics." Mechanism saw mechanics as the leading principle of explanation. The fact that mechanics as a craft had reached a high level before the start of the Copernican revolution contributed significantly to its success. It would be a grave mistake to consider 17th-century mechanical craft the fruit of Copernicanism. Rather, it made Copernicanism and the world view of mechanism successful.

The ontological aspect: cosmology
Any world view has first of all an ontological aspect. It says roughly how the world looks, and also provides the framework into which the solutions of problems have to fit.

It is often said that the shift from the geocentric to the heliocentric world view implies that mankind no longer held the centre of the universe, and had to be content with a more *modest* position.[28] This is typical hindsight. It was by no means the view of Copernicus, Kepler, and Galileo, who knew the background of ancient and medieval cosmology better than our present-day world viewers. In this cosmology, the central position of the earth was by no means considered "important." The earth, including its inhabitants, was considered imperfect, occupying a very low position in the cosmological hierarchy. With the advent of the heliocentric world view, man was not "bereft from his central place", but was "placed into the heavens", the earth becoming a planet at the same level as the "perfect" celestial bodies.[29] To quote Galileo: "As for the earth, we seek rather to enoble and perfect it when we strive to make it like the celestial bodies, and, as it were, place it in the heaven."[30]

The Copernican view of the cosmos was of great influence on the concepts of space and time.[31] In Aristotelian physics space is finite, bounded by the starry sphere, but time is infinite. Aristotle recognized neither beginning nor end of the cosmos and this embarrassed his medieval disciples. The Christian world view requires a beginning, the Creation, as well as an end, the return of Jesus Christ to the earth.

In the Aristotelian view of the cosmos determined by the form-matter motif, the earth stood still at the centre of the universe, which as a whole was not very much larger than the earth. Of course, Aristotle knew that the dimensions of the earth are much smaller than those of the sphere of the stars, but the latter was considered to be small enough to take the argument of stellar parallax seriously (see Sec. 2.4). When Copernicus introduced the annual motion of the earth, he had to enlarge the minimum dimension of the starry sphere such that the distance between the earth and the sun becomes negligible compared to it. This turned out to be a step towards the idea of an infinite universe. Copernicus, Kepler, and Galileo, however, still considered the cosmos to be spatially finite.[32] Descartes, on the other hand, identified physical space with mathematical, Euclidean space, and therefore took it to be infinite.

In Aristotelian physics the immediate environment of an object is its place.[33] The natural place of the element earth is water, surrounding the earth. The natural place of the sphere of fire is above the sphere of air and below the lunar sphere. The place of Saturn is above Jupiter's sphere and below the starry sphere. Descartes agreed that the place of a body is its environment. On the other hand, he realized that the position of a body can be determined with respect to a coordinate system, and is not in need of material surroundings.[34] He vacillated between the views that motion is relative and that it is absolute (see Sec. 3.4).

As we have seen in Sec. 3.6, Newton asserted that the place of a body and its relative motion can be determined with respect to other bodies, eventually a coordinate system. Rectilinear, uniform motion is a relation between two bodies, and no experiment can decide whether one of them is absolutely resting. Nevertheless, Newton held to the idea of an absolute space and time, on religious grounds, sustained by experimental results. If no matter existed, space and time as *"sensoria Dei"* would be determined by God's omnipresence and eternity. Hence, space and time should be infinite.[35] A hundred years later, Immanual Kant would consider space and time as necessary presuppositions of human experience of spatially extended things and temporal events.[36]

The ontological aspect: structure of matter
Aristotle rejected the possibility of a vacuum, on rational grounds (what is not, cannot be); on cosmological grounds (the place of a thing is determined by its surroundings, and a thing in a void would therefore be nowhere); and on physical grounds (the local

motion of a projectile is inversely proportional to the resistance of the medium; hence, in a vacuum, motion would be instantaneous, which is contradicted by sensory experience and by common sense). In the middle ages theories of atoms and the void were always connected to atheism and therefore suspected. In the 17th century Gassendi publicly stated that this allegation was unwarranted. Atomism became such a natural consequence of the rising mechanistic world view that Descartes' aversion to atomism, though he was a mechanist, does not sound consistent with his premises.

Descartes' rejection of atomism has two sources. First, his preoccupation with mathematics (according to which space is infinitely divisible) inhibited the acceptance of indivisible material particles. Second, his identification of matter with space (matter having only one property: extension) excluded the possibility of a void, though on grounds differing from Aristotle's. Nevertheless, Descartes did not escape the spell of atomism, when he postulated the existence of three kinds (or sizes) of material particles.

Galileo considered a vacuum first as an idealized environment in a Platonic sense, in order to study local motion in its purest and unresisted form. His experience with pumps, not being able to elevate water to a greater height than about ten metres, led him to ponder the actual existence of a vacuum. His disciple Torricelli's experiments with a mercury column, later repeated and extended by Pascal and Boyle, and the invention of the air pump, soon convinced the scientific community of the reality of a void. Newton completed this development by showing that the stability of the solar system requires an interplanetary vacuum.

This investigation of the void strongly furthered the atomistic view of matter, though until the end of the 19th century no direct evidence for the existence of atoms was available. Atomism was very influential on the development of chemistry in the 19th century, though it hampered the development of wave optics.

Cartesian and Newtonian views on space and matter are nearly diametrical opposites. Descartes identified space with matter, his world is a plenum, motion can only occur by way of vortices within the plenum, and force is an effect of motion, of collision between material objects, of action by contact. Newton started from the dualism of matter and force, and his world is mostly empty. Force is not the effect, but the cause of change of motion, and he admits action at a distance between bodies as wide apart as the sun from Saturn. Small wonder that at the end of the 17th century, the leading Cartesians, Huygens and Leibniz, both rejected Newton's *Principia*, though both recognized its mathematical superiority. And even after Newton's death, Voltaire (1733) commented: "A Frenchman who arrives in London finds himself in a completely changed world. He left the world *full*; he finds it *empty*. In Paris the universe is composed of vortices of subtle matter; in London there is nothing of that kind. In Paris everything is explained by pressure which nobody understands; in London by attraction which nobody understands either."[37]

The sources of data

A world view also determines the acceptable sources of data. In the middle ages the authority of the Bible, the church, and Aristotle played an important part. During the Copernican revolution the view emerged that natural science has its own sources of data, and is able to judge them on physical rather than on philosophical or theological grounds.

We discussed the sources of data in Sec. 5.3, where we saw the influence of Platonic, Aristotelian, and Copernican world views on the evaluation of data. Plato's idealism led him to overestimate the imagination and the procedure of *anamnesis*. He considered observation deceptive, and like Aristotle he had little use for experiments. Aristotle's realism made him hold common sense knowledge and intuition in high esteem, but he did not value quantitative science. Experiments were useless to achieve knowledge about things, because their artificial character intrudes on the nature of things.

The most obvious deviation of Copernicanism from earlier philosophies is the gradual decline of common sense as a reliable source of knowledge. From Copernicus' statement that the earth moves, via Gilbert's hypothesis that the earth is a magnet, Kepler's laws of planetary motion, and Galileo's principle of inertia, to Newton's assertion that the sun attracts the earth through a vacuum, we perceive an increasing departure from common sense insight. Though Galileo still practiced it, the idea of achieving knowledge through *anamnesis* gradually faded away. Soon, intuition was no longer considered a source of data.

On the other hand, observation and experiment became more and more important. In a mechanistic world view nothing precludes the possibility of achieving knowledge with the help of artificial means such as telescopic observation or experiments with a barometer. In particular the Copernicans stimulated the quantification of science. Measurement became the primary source of data.

This has influenced not only the course of science. The new

view of nature led people to suspect data based on hearsay, on superstition, on unverifiable sources. The conviction that only reproducible data, only measurable or at least observable facts, should be considered reliable, for example, did much to obliterate witch hunting, which was a familiar feature of 16th- and 17th-century life.

Argumentation

Next, the world view determines what kind of arguments are acceptable. Both ancient and medieval scholars, but also Descartes and many 20th-century philosophers highly valued human reasoning, in which logic has a leading part.

With Tycho Brahe, Kepler, Galileo, Boyle and Newton, a different accent appeared. Their reasoning rests more on mathematical analysis than on logic, and observation and experiment are allowed to tip the balance. Pure reasoning or a mathematical deduction is seldom considered sufficient to settle a scientific matter, if it is not sustained by observational and experimental tests.

This was not yet the case with Copernicus himself, but we have seen that Kepler rejected a solution of the problem of the motion of Mars because it deviated by eight minutes from Tycho's careful measurements. As an argument, this would not have counted earlier, but it was a major consideration later. Galileo, Pascal, Boyle, Huygens, and Newton were gifted experimenters.

In other respects, too, the prevailing world view influenced the methods of proof. Huygens' method of solving the problem of impact by a coordinate transformation would have been useless in an Aristotelian context, because it appealed to the principle of relativity. Aristotelianism only knew of motion in an absolute sense, whether natural or violent.

In his *Dialogue*, Galileo described an experiment with plane and convex mirrors hanging on a diffusely reflecting wall, in order to demonstrate the difference between diffuse and specular reflection.[38] He used the result to argue that the surface of the moon cannot be smooth. For him the argument is valid, because in principle he does not recognize a difference between terrestrial and celestial physics, and an experiment performed on earth has consequences even if applied to celestial states of affairs. For an adherent of Aristotle's physics the argument would have little force, because the distinction between sublunary and celestial realms is fundamental to his world view.

Finally, an important change with respect to theories concerns the status of axioms (see Sec. 4.4). In a rationalistic world view, whether Aristotelian or Cartesian, the axioms of a theory should be self-evident, intuitively clear, and in agreement with common sense. Slowly, during the Copernican revolution the insight gained ground that the primary function of scientific theories is not rationally to explain states of affairs starting from evident principles, known to be true, but rather the other way around. The principles of Copernican theories became more and more counter-intuitive. They were intended to explain the known, *i.e.*, the observable world, by the unknown hidden laws of nature.

Heuristics

This leads us to the final part of a scientific world view to be discussed, the heuristic it allows or stimulates. The primary aim of science became the discovery of the lawfulness of nature, not merely by rational thought, but also by observation and experiment.

It is remarkable that the rationalists have hardly any use for the heuristics connected with the four "directions of research" which we identified in Chapter 6. Indeed, if the axioms of science are either self-evident and intuitively understood (as with Aristotle and Descartes), or are found by trial and error (as with Popper), there is hardly any room left for the recognition of heuristics specific for any kind of research. It is therefore not surprising that these rationalists also tend to hypostatize a single mode of experience, for instance, the physical one (Popper), or the kinematic one (Descartes).[39]

Only a realization of the possibilities *given* to us in nature makes room for different heuristics. Even then, it is possible to hypostatize one of them at the cost of others. Thus the Copernicans particularly favoured mathematization as a heuristic, but also abstraction as a means to finding the universal laws of motion and of physical interaction. Realists favour the method of successive approximation in the investigation of the structure of matter, and pragmatists propagate the application of science as the main drive for its development.

12.4. The hypostatization of theoretical thought

In Sec. 1.3 we observed that the logical structure of a theory precludes the possibility that a theory can prove a statement conclusively. Nevertheless, many philosophers have tried to hypostatize theoretical thought. Because theories are man-made, hypostatization means that man tries to constitute reality by his theories. For Aristotle this meant the identification of God (the prime mover) with theoretical thought. For Descartes the hypostatization is guaranteed by God. Descartes is the founder of the humanist philosophy of "nature and freedom." For Kant, man is constrained by certain categories of understanding in his theoretical reconstruction of the world. Kuhn today asserts that reality is structured by a paradigm.

We shall briefly discuss the views of Aristotle and Descartes before criticizing the idea of hypostatization.

Aristotle

In Sec. 3.2 we summarized Aristotle's scheme of explanation. Together with his cosmology it constitutes a hierarchically ordered universe, in which all change means the actualization of the potential properties of matter towards the realization of its form. Strictly speaking, Aristotle was an atheist, for his philosophy excluded a personal God. But the forms themselves have a divine nature — they are eternal, perfect, unchangeable, and understandable. All change is determined by a final cause, so as to achieve a perfect form. God is immanent in the world as its intelligible order, and transcends the world as its ideal end.[40] But this god is nothing but pure thought, and being perfect, it can only think about something perfect, namely, itself. It is the prime mover, because everything strives to become perfect, to become like god. "The only reason for the existence of the world as a whole, the only fact that makes it more than a sheer occurrence, and renders it intelligible, the only fact that justifies nature to man, is that the world exists to make life possible, and at its fullest, to make possible the best life, which for Aristotle is the life of sheer knowing, "*nous* nousing *nous*.'"[41]

Descartes

Both being rationalists, Aristotle and Descartes have more in common than Descartes would be ready to admit. Apparently, there are several important distinctions. Descartes always stressed that he was faithful to the Catholic religion, to the faith of his youth.[42] However, when in his philosophy he speaks about God, one cannot escape the feeling that his God does not differ very much from Aristotle's.

Again, Descartes rejects the Aristotelian form-matter scheme and its associated principle of explanation. As a mechanist he rejects the use of final causes. Everything must be explained in terms of matter and motion.

Descartes was very careful not to offend the church. Adopting an instrumentalist position, he asserted that God could have made the world such that two plus two makes five, but in that case it would be unintelligible. In order to avoid giving offence, he says that the world he describes is not necessarily the real world, but a possible world, a rationally conceivable world.[43] It is to be noted that behind this pious caution, Descartes hides his real objective, which is *to create a world according to his theories*. If this world differs from the observed world, the latter should be pitied, and anyhow, observations are deceptive according to Descartes.[44]

Descartes' philosophy is well-known because of his "methodical doubt." Descartes conquered his doubt by showing that he cannot doubt his own existence. His *"Cogito ergo sum"* became the hallmark of his philosophy.[45] But by doubting, by being uncertain, man is imperfect. This conclusion requires that he cannot doubt the existence of a perfect being, God.[46] But God, being perfect, will not deceive us whenever we have a "clear and distinct idea."[47] God warrants the existence of anything which we clearly and distinctly perceive.[48]

This is a *theory*. It means that Descartes attempts to prove theoretically that he himself exists, that God exists, and that the clear and distinct ideas, which form the axioms of other theories, are true. During his lifetime, Descartes was criticized for the circularity in his reasoning. It seemed that Descartes believed in God because he had a clear and distinct idea of him, and that the clear and distinct ideas are true because they are warranted by God. The circularity is less obvious if we compare Descartes' God with Aristotle's prime mover. It is nothing but the hypostatization of theoretical thought, and later atheist philosophers had no trouble adhering to Cartesianism.

Of course, this reasoning falters if it would depend on Descartes' subjective doubt. Therefore, he opens his *Discourse on Method* (1637) with the statement: "Good sense is mankind's most equitable divided endowment."[49] Only if his methodical doubt has a universal character, and if everybody agrees with his train of thought, is it able to constitute the world.

But man can only construct the world if he is free, if he stands opposite nature. On the other hand, man is part of nature. For this reason, Descartes makes a sharp division between body and mind. "Reason is the only thing which makes us men and distinguishes us from the animals."[50] Nature, or *res extensa* (extended being), including the human body, is determined by natural laws, whereas the human mind, or *res cogitans* (thinking being) is free to constitute the world according to reason. Descartes is more certain about his thought, his mind, than about his body.[51]

Hence Descartes arrives at the insight that there are "clear and distinct" ideas whose truth is warranted by God.[52] He calls "clear" whatever is present and manifest to the mind, and "distinct" whatever is precise and different from everything else.[53] Their truth rests on intuitive insight rather than on observation or authority. Self-evident truth is so clear and distinct that nobody can doubt it.

Descartes convinced himself and his followers that the true first principles of physics are *matter*, spatially defined by size and shape, and *motion*. All physical phenomena can only be rationally explained from these principles. In particular, the laws of motion, its conservation and rectilinearity, and the laws of impact (Sec. 3.4) are true laws of nature, beyond any doubt.[54] The most general parts of his physics were as certain as the mathematical truths that two plus three makes five, and that the sum of the angles of a triangle is twice a right angle. (He could have known better had he studied spherical geometry.)

However, as soon as Descartes tried to explain particular phenomena, he had to admit that his hypotheses did not carry the same certainty as his first principles (see Sec. 11.3). There are so many ways to deduce the properties of light, for instance, that he could never be certain that he had given the true account. Only then would Descartes admit observation and experiment as the means of deciding between various possibilities. However, no possible explanation should be admitted if it contradicted the true first principles of his philosophy, *i.e.*, the mechanical principles of matter and motion.

Popper's transcendental critique
According to Kant, Descartes was uncritical with respect to his hypostatization of theoretical thought. Kant introduced the idea of "transcendental critique", to be distinguished from "immanent" and "transcendent" criticism (Sec. 10.2).[55] Popper also accepts this idea. It can hardly be denied that his distinguishing among "three worlds" is reminiscent of Descartes' distinction between *res extensa* (Popper's first world of physical states) and *res cogitans* (Popper's second world of mental states). The introduction of a third world (of ideas in the objective sense, of theories, arguments and problem situations, see Sec. 1.2) is only a refinement, though a useful one. However, Popper does not introduce a *deus ex machina*, and he is more critical than Descartes.

The transcendental method accepts scientific knowledge as a fact, and questions the principles according to which this fact can be explained.[56] According to Popper transcendental critique can only be used in a negative sense, namely, to criticize certain opinions. In this way, he rejects any inductivist theory, which leads to a denial of the hypothetical-deductive method, and considers theories to be superfluous.[57] By this double negation Popper obtains a confirmation of the hypothetical-deductive method, of course. But this confirmation cannot be based on the experience that this method is fruitful, because that would be an inductive argument. Similarly, it cannot be proved by the hypothetical-deductive method, because that would be a *petitio principii*. Likewise, the inductive method cannot be justified by induction. It remains that Popper *believes* his method, like the inductivists believe theirs.

In this way transcendental critique must be applied. It probably does not convince anybody, but it reveals the dynamic motifs of the philosopher concerned. It lays bare the ultimate *beliefs* of the philosophy or world view to be scrutinized.

One may wonder whether the transcendental method is applicable to Popper. Transcendental criticism questions the *certainty of theoretical thought*, and Popper denies the possibility of certain knowledge. It is impossible to confirm the truth of hypotheses or theories, although it is sometimes possible to falsify them. Hence, the problem is not to find the truth of a statement, but to decide which theory should be preferred. The answer to this problem is that we should prefer those theories which have withstood criticism better than their rivals.[58] This does not mean that Popper relativizes the question of truth. The classical idea of absolute or objective truth is accepted as a *regulative idea*, a standard of which we may fall short.[59] If we reject a theory in favour of another one, we do so because we think the latter comes closer to the truth, but we can never be certain about this conjecture.

However, Popper also has his certainties. He is convinced of his hypothetical-deductive method, of logic, of realism, of objective truth, of the existence of natural laws, and of the meaningfulness of rational criticism. Popper admits this, and he calls it his *metaphysics*, because these certainties concern metatheoretical views. The impossibility of acquiring certain knowledge only concerns theoretical knowledge of the first world, achieved by means of theories. But Popper is convinced of his metaphysics.

Popper's realism includes the metaphysical view that natural laws exist independent of mankind, unchangeable, always and everywhere valid.[60] As a statement, this belief is part of a metaphysical cosmology.[61] The idea of law is inborn, and the fallible method by which one finds law statements is the method of trial and error.

The origin of natural laws, which Descartes ascribes to God, is according to Popper an unsolvable problem in his metaphysical realism.[62] The problem concerning the universal validity of natural laws makes Popper think of Newton's concepts of space and time as *sensoria Dei*. The immediate action at a distance of the gravitational force reminds one of God's omnipresence. Popper recognizes the same problem with respect to the universal validity of natural laws. This cannot be explained by interaction, because interaction is explained by the laws.[63] We have to accept the existence of natural laws as a mystery, Popper concludes.[64]

Dooyeweerd's transcendental critique
In his *A New Critique of Theoretical Thought* the Dutch Christian philosopher Herman Dooyeweerd attempts to lay bare the basic presuppositions of any philosophy. He was mainly concerned with Kant, but his critique can also be applied to Popper's views, as we shall briefly sketch.

Dooyeweerd's transcendental critique of theoretical thought proceeds through several phases.[65] The first phase concerns the question of the structure of theoretical thought. We have answered this question by considering theories, statements, and concepts to be instruments of thought, acting between thinking subjects and objects of thought. As we observed, this looks like Popper's three worlds. By the introduction of these logical artifacts, reality is taken apart because it leads one to an antithesis between logical subject and logical object, which is absent in natural thought (Sec. 1.2).

This leads to the second phase of Dooyeweerd's transcendental critique, namely, the question of how one arrives at a synthesis between those parts of reality which theoretical thought takes apart. Popper's answer was found in his three-worlds theory. The synthesis between worlds 1 and 2 seems to be obtained in world 3. The ideas and theories are free inventions of the human mind, but they are tested with the help of world 1 objects.

Dooyeweerd observes that the synthesis only becomes apparent by way of critical self-reflection, because the human subject is involved. Similarly, Popper says man can transcend himself in his theoretical activity.[66] By inventing theories, man attains an intellectual liberation.[67]

Hence, the third phase of transcendental critique is to ask how this critical self-reflection is possible. Popper's answer is: by the method of rational criticism, that is, by trial and error, by learning from one's mistakes. By this process man is able to transcend himself. This method also determines the community of thought, because Popper invites everybody to criticize him.

However, such a community can only be constituted by a communal belief, a religious commitment or an ideology determining our thought and activity, a common world view. This leads to the need for regulative ideas — the ideas of origin, unity, and coherence of reality, which direct theoretical thought. It is impossible to derive theoretical knowledge from these ideas, but the method of theoretical research is determined by them. Popper recognizes the existence and value of regulative ideas, such as the idea of absolute truth, even if he calls them metaphysical rather than religious. Popper's dynamic motif does not really differ from Descartes' motif of nature and human freedom.

Popper seeks a synthesis of nature and freedom in his world 3, where people freely invent their hypotheses, in order to test them in world 1. By this synthesis man liberates himself. By rational criticism man liberates himself from uncritically accepted prejudices — not from all prejudices, however. In any case, Popper does not transcend *logic*, which is tautological, hence not subject to criticism. He also uncritically accepts the idea that logic and mathematics are not empirical sciences (see Sec. 10.1). Hence, autonomous thought remains partly "original" in the strict sense of this word, it is not subject to criticism.

The coherence of the three worlds is weak. Popper suggests the existence of an interaction between worlds 1 and 2, and also between worlds 2 and 3. Within world 1 there is a continuous coherence of things, plants, animals, and so on, and within world 3 there is a similar continuity of ideas. Hence the idea of mutually irreducible modes of experience, discussed in Chapter 3, will not be found in Popper's ideology. It is eradicated by the continuity of absolutized thought.

Popper distinguishes only between testable, hence empirical ideas, and non-testable ideas taken from logic, mathematics, and metaphysics. The demarcation between testable and non-testable ideas is found within world 3, and is not bridged. Therefore, the dualism of nature and freedom manifests itself even in world 3, meaning that world 3 cannot provide a synthesis between worlds 1 and 2. Thus Popper does not succeed in bridging the duality introduced by Descartes, the duality of *res cogitans* and *res extensa*. It means that rational criticism is unable to transcend the dualism of nature and freedom, and cannot liberate man from this prejudice. Hence, Popper's idea of self-transcendence is an illusion.

12.5. The unity of the creation

Dooyeweerd's transcendental critique does not reject theoretical thought as such, but its pretended autonomy, its absolutization or hypostatization. The unity of the world should not be *given* by theoretical thought, but must be considered its presupposition, transcending any theory.

The knowledge of God and self-knowledge

One can hardly avoid comparing the idea of self-transcendence with the famous Baron of Muenchhausen's attempt to draw himself out of the morass by pulling at his wig. Or, to put it less frivolously, self-knowledge is impossible without revelation. Calvin stressed that true self-knowledge is only possible by, and nearly identical to, knowledge of God. But Calvin did not mean a theoretical idea of God as Descartes did. Calvin referred to God, the father of Jesus Christ, who reveals him to us. Descartes' God was subject to logical laws, but Calvin asserted that not only natural laws but even the laws of logic hold true only as long as they are maintained by the Creator, because of his covenant in which Jesus Christ is the mediator. As we have seen, this reformational view opens the possibility of rejecting rationalism, and of investigating the lawful structure of the world with the help of logical instruments like theories.

But this view also cuts off the idea of theoretical self-knowledge. Self-knowledge is a fruit of religion and can be true or false, not according to the criteria of some theory, but because the religious premise is true or false. According to Calvinism, man is completely dependent on his Creator. He is allowed to investigate the creation, but should never presume to become its master, whether in thought or in any other activity.

Theories can be used to obtain knowledge of laws, but cannot help us to get knowledge of God. Only because he sent Jesus Christ into the world to become subject to the laws, has God made himself known to us.

The principle of excluded antinomies

According to Descartes, the truth of "clear and distinct" ideas is

guaranteed by God's perfection. Although our senses are deceptive, God will not deceive us. We have criticized this view, because it makes God a rational being, a logical subject, instead of
the supreme lawgiver, even of logical laws.

Nevertheless, it cannot be denied that the Calvinist viewpoint, too, contains something similar to Descartes' warranty. We
have seen that, as logically qualified artifacts, theories are subject to the logical law of non-contradiction. One of the main functions of theories is to connect statements, and thus to systematize
our knowledge. We use theories to compare statements with each
other. For practical reasons we restrict the statements to be compared to a limited set, what we have called the "context of the
theory." *Within* this context, a theory must be free from contradictions. If only temporally and tentatively, a theory is allowed to
have contradictions *outside* this context.

But it is clear that with respect to the most fundamental statements, in particular law statements, contradictions cannot be accepted in the long run. Now the principle of non-contradiction is a *logical* principle. But if we assert that law statements ought not to contradict each other, this logical norm refers to an *ontological* state
of affairs. It is assumed that laws are consistent with each other,
in an ontological sense. This is the *principle of excluded antinomies* (*nomos* is "law" in Greek).[68] It cannot be proved, and should
not be proved by assuming God to be consistent, or rational. It can
only be believed. It is an expression of the belief in the cosmic
order.

Hence, we reinforce our conclusion that ultimate certainty cannot be obtained by way of theories. Whatever absolute certainty
we have is a matter of belief. But it cannot be denied that the use
of scientific theories gives us more certainty (in a relative sense)
about the laws of nature than any other method known. We cannot
but be content with this.

The acceptance of the *cosmological* principle of excluded antinomies allows us to investigate the world in a piecemeal manner.
There is no need to worry constantly whether a new theory fits
into the network of existing theories, if we believe that the logical principle of non-contradiction is subordinate to the cosmological principle of excluded antinomy. Because the latter is valid independent of our thought, we can tentatively and temporally
afford to contemplate theories which contradict other accepted
theories, presuppositions, world views, and data. We can afford
to restrict temporally the principle of non-contradiction to the
theory under investigation. Thus, Copernicus could afford to forget

about Aristotelian physics, as long as his theory was internally consistent and fitted the facts which he chose to take into consideration.

For a Christian, the principle of excluded antinomies is a consequence of his belief in the unity, coherence and diversity of the creation, warranted by God, the Creator of heaven and earth. This implies the rejection of any absolutization of some aspect or part of the creation, because this would take the place of God. The fundamental unity of everything transcends the creation, and one can only achieve an idea of its existence by one's belief in God. The rejection of any kind of hypostatization could liberate us from the need to find a single authoritative viewpoint from which to interpret anything. We are then free to take any viewpoint, to change our viewpoint, to tackle our problems in any way.

Responsible belief

The statement "the earth moves" is the fruit of a theory which cannot be proved in an absolute sense. There, of course, are sound objective reasons for accepting this theory, enabling us to describe it as well-corroborated. Nevertheless it is highly remarkable that all civilized people *believe* that the earth moves. They believe this on the authority of science, though science professes to reject authority as a source of knowledge. Most people do not know which objective arguments are put forward to argue in favour of the Copernican thesis, and if they know they do not understand them or are unable to have a critical opinion of the theories from which it follows. This means that science, after being emancipated from the authority of philosophy or theology, has assumed an authority of its own. This would not be too bad as long as it were restricted to purely scientific matters. However, in our time science has achieved an authority in education, in business, in politics, and in everyday life, which is hardly matched by any other influence in our present or past culture.

Since the 19th century many people turning their back on Christian belief have taken science to be the true and authoritative source of knowledge. Yet, for all its marvellous achievements, scientific theoretical thought, if absolutized, leads to an alienation between human beings and the creational order. The more science seems to be able to master the world, the more people feel their freedom threatened.

This alienation can only be overcome by true self-knowledge, leading to the recognition that the logical relation between human, thinking subjects and the objects of their thought, instrumentally separated by theories, is not rigid, is not unique, and is not a necessary requisite to view our world.

We have emphasized that theories, starting from counter-intuitive axioms, can be used to criticize common sense. This should not deceive us into believing that human intuition is useless. It would be useless if theoretical thought were allowed to determine our life totally. But mankind should fight any kind of totalitarian rule, whether by the church, the state, or by science. It is an intuitive but reliable idea, a true belief, that totalitarian rule destroys human responsibility.

By using theories we are able to objectify *part* of the world, *any* part — but never the world as a whole, because we are part of it. On the other hand, because we are part of the world, even before beginning to think about it, we are aware of the laws to which the world is subject. This natural order is not *made* by theoretical thought, but is in part discovered by it. Before we discover it, we are intuitively aware of the natural order, because we are subject to it ourselves. Therefore, theoretical thought takes its starting point in intuition, even though it eventually criticizes it.

In particular, theoretical thought can never prove the existence of our "self", our own existence. As we have seen, Descartes tried to deliver such a proof, but in vain. Nor is theoretical thought able to prove the existence of laws. In fact, there is no need of either proof. We know that we exist, and we are aware of the existence of laws. Otherwise, theoretical thought would be senseless.

In its most pregnant sense, human responsibility means to be responsible to one's creator. The 16th-century reformers stressed that every man and every woman is responsible for his or her own beliefs and deeds. Everyone has to respond to his calling by God. This means the rejection of any kind of hypostatization, of any kind of totalitarian rule.

It also means that we cannot relegate the problem of truth to theories. Theories are merely *instruments* of thought, and only scientists and other users of theories can be held responsible for their use. Hence we cannot find our certainty or place our trust in theories. Certainty is unbreakably connected with personal belief. It is up to every man or woman to decide for him- or herself what to believe.

The hypostatization of theoretical thought means the attempt to incorporate everything into a theoretical framework. It inevitably leads to the division of the creation into human, subjective freedom and subhuman, objective nature, into *res cogitans* and *res extensa*, into mind and body, or into Popper's somewhat more sophisticated "worlds." However, if we recognize theoretical thought to be of limited relevance, to be concerned with only the logical aspect of human experience, we may be able to incorporate theoretical thought into the wholeness of our experience, which unity is not warranted by theoretical thought, but by our beliefs.

Notes

Chapter 1 : **The logic of theories**
1. For our bibliography, see the Index. For biographical details, see in particular Gillispie (ed.).
2. Kant (a) B xiv, xxii; but see Cohen (f) 237-253
3. Kuhn (a) 1, 2; Brown 111; Dijksterhuis 317; Toulmin, Goodfield 179. But there is little agreement about the term "revolution" with respect to the history of science, see Cohen (e) 3
4. Koestler, part IV
5. On positivism, see Kolakowski; Von Mises; Suppe
6. See, *e.g.,* Suppe; Brown; Glymour, Ch.2,3; Weimer
7. The artificial character of science is stressed by Ravetz, Ch.3, 4, and by Polanyi, Ch.1, 4
8. Copernicus 24 (Preface)
9. Popper (d) 154
10. The "theory of theories" developed in this book is based on Herman Dooyeweerd's philosophy. Cf. Stafleu (e)
11. Popper (f) 33 identifies a theory with a hypothesis, but on p.113, 178, 292, with a deductive system. Cf. Popper (a) 59: "Scientific theories are universal statements."
12. cf. Braithwaite 12, 22; Bunge (a) 51-54; (c) 381
13. The term "statement" should not be taken too restrictively. It includes mathematical equations, tables, graphs, *etc.*
14. Bunge (c) 391. A theory is "closed" with respect to deduction, but is quite open in another respect, as we shall see.
15. See *e.g.,* Suppes; Cohen, Nagel
16. L.E.J. Brouwer and other intuitionists accepted a proof only if it is finite. Proof by complete induction is therefore rejected. Brouwer also rejected *reductio ad absurdum* as proof
17. cf. Finocchiaro (b) 311-331
18. Popper (b) 317-322. Formally: $p \to (p \vee q)$; $(p \vee q, -p) \to q$; hence $(p \wedge -p) \to q$, whatever q
19. Heath 299-310; Dreyer 136-148
20. Copernicus, *Commentariolus;* see Rosen (a) 59

21. Leibniz considered two objects to be "identical" if they cannot be discerned from each other, if they are equivalent in any sense; see Broad 39-43. This view is now refuted by the existence of photons and other "bosons", which can be fully indistinguishable, and yet have to be considered individual particles — e.g., the number of indistinguishable photons in a certain quantum state can be determined

22. Randall 136-137

23. Galileo (a) 369; Descartes (a) 20

24. Bridgman 5. For a criticism, see Hempel (a) 39-50; (b) 123-133, 141-146; (c) Ch.7; Bunge (b)

25. Newton (a) 1; cf. Jammer (c) 64-74; Cohen (b) 335

26. Mach 237, 300

27. For an English translation of La Bilancetta, see Fermi, Bernardini 134-140

28. Drake (ed.) 79-81

29. Newton (a) 414

30. Or see Suppes; Cohen, Nagel

31. Popper (b) Ch. 11; (f) 194-216, 261-278; Kolakowski 212-216; Von Mises, Ch.6

32. Extensional logic is restricted to the extension of concepts; predicate logic also concerns their intension. See Bunge (c) 65-72 on extension and intension

33. Galileo, Sidereus Nuncius, in Drake (ed.) 57; Galileo (a) 334, 339-340. This discovery also showed that celestial bodies can partake in two motions simultaneously

34. Kuhn (b) Ch.9,10; Feyerabend (a); Hesse (b) 61-66; Suppe 199-208; Stafleu (a) 25-27.

35. Feyerabend (e); Hanson (a) 5-9, 18-19; cf. Brown, Ch.6

36. Finocchiaro (b) 311: "An argument is a basic unit of reasoning in the sense that it is a piece of reasoning sufficiently self-contained as to constitute by itself a more or less autonomous instance of reasoning."

37. cf. Polanyi; Cantore

Chapter 2: **Explanation and prediction**

1. Influenced by Mach, most logical-empiricists adhered to instrumentalism. It has been criticized by Popper (b) Ch.3; (f) 111-149; Bunge (a); Feyerabend (b); Giedymin

2. Plato 1164-1165 (Timaeus); cf. Dreyer, Ch.3; Heath, Ch.15

3. Dreyer, Ch.4, 5; Heath, Ch.16

4. Aristotle's system counted 55 spheres, six more than was necessary. Cf. Aristotle (a) XII,8

4a. Duhem (b) 5

5. cf. Kuhn (a) 48

6. For a detailed discussion of Ptolemy's work see Neugebauer (a), (b); Dreyer, Ch.9

7. Koestler 209; Rosen (b), Ch.3

8. Dijksterhuis 230-237; Duhem (b), Ch.2-4

9. Copernicus 27 (Preface)

10. Kuhn's (a) 125, 196 statement that Copernicus' work was used applies to the calculations, not to the principle of the moving earth

11. Plato 1161 ff (*Timaeus*). Copernicus 25 (Preface) refers to this craftsman: "...the mechanism of the universe which has been established for us by the best and most systematic craftsman of all..." It became a recurrent theme in 17th-century mechanism

12. Dijksterhuis 237-241, 254-256; Hooykaas (a) 75-79; Kuhn (a) 114-122; Toulmin, Goodfield 165-169; North. Buridan and Oresme were the most important representatives of the "Paris Terminists", see Dijksterhuis 181-229; Clavelin, Ch.2; Hooykaas (a) 75-85

13. Hooykaas (a) 77-79

14. Dijksterhuis 185,186; Giedymin

15. Copernicus' *Revolutionibus* was published in 1543 (Nürnberg), 1566 (Frankfurt) and 1617 (Amsterdam)

16. Copernicus' *Commentariolus* or *Sketch of his Hypotheses for the Heavenly Motions*, was never published in print during the Copernican revolution. For an English translation, see Rosen (a) 57-90. Rheticus' *Narratio Prima* (Rosen (a) 107-196) was published 1540, 1541, 1566 (together with Copernicus' *Revolutionibus*), 1597 (together with Kepler's *Mysterium Cosmographicum*) and 1621 (again). It was intended to be followed up by a *Narratio Secunda* (Second Story), which was never written, see Rosen (a) 162, 186

17. cf. Koyré (a) 3, 36, 201, 202

18. Copernicus 25 (Preface)

19. Rheticus, *Narratio Prima*, Rosen (a) 166, 135, 137; Copernicus, *Commentariolus*, Rosen (a) 59; cf. Galileo (a) 341-342

20. Copernicus 25 (Preface)

21. *ibid.*

22. Galileo (a) 122

23. Galileo (b) 96, 97

24. *ibid.*, Fourth Day

25. cf. Bunge (c) 354-355

26. Kepler (b) 34 (Introduction), 267 (Ch.44), 345 (Ch.58); cf. Koyré (c) 225, 244, 264

27. Kepler (b) 24 (Introduction), 247 (Ch.40); cf. Koyré (c) 234

28. Galileo (a) 398-399, 410-411
29. Kepler (b). *Astronomia Nova* was published in 1609, and not reprinted before 1800. Written in Latin, the only translation into modern languages is the German one. This is also the case with Kepler (c), *Harmonice Mundi* (1619)
30. Kepler (c) 291 (book V, Ch.3)
31. Koestler 204. The reference to "ellipses" can only be found in the manuscript of *Revolutionibus*, where it is crossed out
32. cf. Finocchiaro (b) 275
33. Bunge (d) 9, 10
34. Galileo (a) 234
35. Popper (a) 69: "...natural laws...insist on the non-existence of certain things or states of affairs..."
36. Popper (b) 33-41; (e) 23-29. For a critique, see Grünbaum (b)
37. Popper (b) 81-83; cf. Hooykaas (a) 23, 223
38. cf. Popper (a) 34-42, 311-314; (b) Ch.11
39. Popper (b) Ch.11; (f) 159-193
40. Copernicus 38-40 (book I, Ch.4)
41. *ibid.*, 46 (I,9)
42. *ibid.*, 37-38 (I,3)
43. Feyerabend (e) 109-111. Ptolemy did not use this coincidence to explain the variation in brightness, because such an explanation would have given a completely wrong value for the variation of the apparent magnitude of the moon and of Venus
44. Copernicus 51 (I,10); 238-242 (V,3); 291-294 (V,35); cf. Kepler (a) 30, 36; Galileo (a) 342-345; Koyré (c)129; Glymour 178-203
45. Galileo (a) 322
46. Copernicus 26 (Preface)
47. Kuhn (a)180 seems to miss this point entirely, saying: "Copernicus' arguments are not pragmatic. They appeal, if at all, not to the utilitarian sense of the predicting astronomer, but to his aesthetic sense and to that alone."
48. Lakatos (b) 189
49. Lakatos (b) 115 gives a wrong impression saying: "...Copernicans *predicted* the phases of Venus, while the Tychonians only explained them by *post hoc* adjustments." On the phases of Venus, see also Secs. 4.3 and 11.1
50. Popper (a) 82-84
51. Aristotle (c) II,14 rejects the motion of the earth using the stellar parallax as an argument. Galileo (a) 138 calls it the most subtle argument against the Copernican position which can be found, cf. Galileo (a) 372-389
52. Hempel (b) 249, 366-376; Carnap 16. For a critique, see Hanson

(c) 161. Hanson's unfinished book was intended to demonstrate the distinction between explanation and prediction, in particular with respect to planetary motion, cf. Finocchiaro (a) Ch.2; Radnitzky; Toulmin (a)

53. Tycho calculated that if Copernicus were right, the brightest stars would be as large as the earth's orbit around the sun. In 1632, Galileo proved Tycho to be wrong, cf. Galileo (a) 358-361

54. Koyré (a) 141-143

55. Dijksterhuis 322-323. Rheticus' *Narratio Prima* does not mention this.

56. Copernicus 47, 48 (I,10) stresses the arbitrariness of Ptolemy's and others' order of the planetary orbits

57. Koyré (c) 51. But it is overlooked by Blumenberg 279

58. Copernicus 51 (I,10)

59. Copernicus' values for Saturn and Mercury differ less than 4%, for Mars, Venus and Jupiter less than 1% from modern values, cf. Copernicus 254 (V,9), 260 (V,14), 266 (V,19), 268 (V,21), 276 (V,27); Koyré (c) 108

60. Kepler (a). The book carries the date of 1596, but was published in 1597. Kepler republished it in 1621, with many additional notes

61. Burtt 64; Rheticus, *Narratio Prima*, Rosen (a) 147, says: "Who could have chosen a more suitable and more appropriate number than six?", but Kepler (a) 21, 26, 27 (Preface) criticizes this view

62. Kepler (a) 23, 24 (Preface)

63. A proof is given in Euclid's *Elements*

64. Kepler (a) 50 (Ch.2)

65. *ibid.*, 89, 92 (Ch.13,14)

66. Galileo (a) 29, and Drake's note to this page, p.470

67. Galileo, *Letters on the Sunspots*, Drake (ed.) 107-109, Galileo (a) 345-356. Galileo explains that according to the heliocentric system the sun has only one motion: rotation around its own axis in 30 days. If one assumes the earth to be at rest, one has to ascribe two more motions to the sun: the daily and annual motion around the earth. The daily motion of the sun would imply that the direction of the axis of the sun's own rotation changes continually, which is dynamically hard to believe. See Drake's note on page 486 to Galileo (a) 354.

Chapter 3: Principles of explanation

1. The idea of "fundamental modes of experience" or "modal aspects" is due to Herman Dooyeweerd. For a systematic analysis of the kinematic and physical aspects, in particular with respect to

modern physics, see Stafleu (a)

2. Initially the Pythagorean universe was geocentric. Sometimes the sphere of the stars was included, sometimes the central fire, sometimes both. In the latter case, the counter-earth was superfluous. See Dreyer, Ch.2; Heath Ch.6, 12; Guthrie vol.1, 282-301

3. Kepler (c); Koyré (c) 326-343

4. Plato 1179-1186 (*Timaeus*)

5. Popper (b) Ch.2

6. The pseudo-Aristotelian *Questions of Mechanics*, extensively annotated by Galileo, only became available after 1525; see Clavelin, Ch.1

7. Aristotle (b) IV,8

8. Toulmin, Goodfield 99 compare this law with Stokes' law concerning the speed of a body moving through a resistive medium

9. Koyré (a) 41; Clavelin 57-58

10. Koyré (a) 7

11. Aristotle on projectile motion: (b) VIII,10; cf. Koyré (a) 51. Galileo was the first to recognize that the trajectory of a projectile is curved right from the start. For his critique of Aristotle's theory of projectile motion, see Galileo (a) 151

12. Drake (c) 38

13. cf. Clavelin 96-97. The latter was Aristotle's opinion, cf. Aristotle (c) I,8

14. Aristotle (b) I,7

15. *ibid.* V,2

16. *ibid.* I,8, 9

17. *ibid.* III,1. By the distinction between eternal form and matter, and changeable substance with its potential and actual properties, Aristotle avoided Parmenides' problem

18. Aristotle (a) I,3, V,2; (b) II,3,7

19. Aristotle (a) XII,2; (b) III,1. In (b) V,1 and elsewhere Aristotle only mentions three kinds of motion, because the first, generation and corruption, is not treated in his *Physics*

20. Aristotle (b) VIII,7,9

21. *ibid.* VIII,8,9

22. Aristotle (c) I,2,3

23. Galileo, *Il Saggiatore*, Drake (ed.) 276-277. Taste, odour, sound, touch and sight are connected, respectively, with water (fluids), fire, air, earth, and aether. Cf. Plato 1186-1192 (*Timaeus*)

24. cf. Salmon (ed.) 5-16, 45-58. The main source of Zeno's paradoxes is Aristotle (b) VI,2,9, VIII,8; cf. Clavelin 34-48; Guthrie vol.2, 91-96

25. cf. Plato's legend of the cave, Plato 747-750 (*Republic* VII)

26. I am not aware of any Copernican who actually discussed Zeno's paradoxes

27. Drake (b) 42. It is a matter of dispute whether Galileo's study of motion was inspired by Copernicanism. Although he openly adhered to it only after 1609, it appears that he accepted the Copernican system after about 1590, see Drake (a); Dijksterhuis 372; Clavelin 177. We shall refer to Galileo (a) by *"Dialogue"* and to Galileo (b) by *"Discorsi,"* although the English title of Galileo (b) reads "Dialogues" for "Discorsi."

28. Galileo (a) 71-78

29. *ibid.* 54, 58; Galileo, *Letters on the Sunspots*, Drake (ed.) 98

30. Galileo (a) 51

31. *ibid.* 412

32. Koyré (a) 130, 131; Descartes (c) 77, 78. In this respect, Copernicanism was preceded by the 14th-century Paris scholars, cf. Dijksterhuis 193-194

33. Galileo (a) 116: Motion does not act

34. *ibid.* 21: Rest is an infinite degree of slowness

35. Galileo, *Letters on the Sunspots*, Drake (ed.) 113-114

36. Copernicus 38-40 (I,4)

37. Galileo (a) 19; (b) 215; *Letters on the Sunspots*, Drake (ed.) 113

38. Galileo (b) 161; Koyré (a) 181

39. Koyré (a) 180

40. Galileo (a) 145-148

41. *ibid.* 20-21; Galileo (b) 261

42. Galileo (b) 264-269

43. Galileo (a) 175. A composite motion is as natural as a simple motion, see Galileo (a) 235.

44. *ibid.*, 398

45. Galileo (b) 276. Tartaglia mentioned this fact in 1531, see Drake (c) 26

46. Galileo (b) 98-99; see Drake (c) Ch.2

47. In his *Letters on the Sunspots*, Drake (ed.) 97, Galileo writes: "...the true constitution of the universe...exists; it is unique, true, real, and could not possibly be otherwise...", cf. his *Letter to the Grand Duchess Christina*, Drake (ed.) 166; cf. Kolakowski 28, 29; Dijksterhuis 372-374

48. Galileo (a) 234, 235

49. Koyré (a) 26, 27, 31; cf. Galileo (b) 62

50. Galileo (b) 72-84

51. *ibid.* 178-179; cf. Drake (b) 84-104, 123-125

52. Aristotle (c) I,8

53. Galileo (b) 167; Koyré (a) 65 ff; Hanson (a) 37 ff, 89; Finocchiaro (a) 86 ff. The same mistake was made by Albert of Saxony, see Clavelin 99, and later by Descartes. Beeckman, misunderstanding Descartes' proof, arrived at the right law of fall

54. Galileo (b) 74; see Drake (c) 39, 40

55. Galileo (b) 174

56. Galileo (a) 221-222; (b) 153, 175

57. Galileo (a) 125-133 mentions all arguments against terrestrial motion, and (133-218) refutes them. Cf. Finocchiaro (b) 208; Copernicus 42-46 (I,7,8)

58. In his *Discorsi*, Galileo does not apply the principle of relativity. To use this principle to refute arguments against the earth's motion is something quite different from applying it in a mathematical theory, as Huygens did, several decades later.

59. Galileo (a) 274

60. Galileo (a) 416-465; cf. Finocchiaro (b) 16-18

61. Kepler (b) 26, 27 (Preface); cf. Koyré (c) 194; Galileo (a) 462

62. Galileo's theory of tidal motion probably dates from about 1595, see Drake (b) 36-44. Galileo developed this theory in 1616. It circulated as a manuscript, *Discorsi sopra il flusso e reflusso del mare*. In 1619, Francis Bacon rejected Galileo's theory because of his (Bacon's) observations, published in 1616. Galileo rejected the observations as far as contradicting his theory. Nevertheless, Shea's (186) characterization of Galileo's theory as "a skeleton in the cupboard" is wide of the mark

63. Newton (a) 435-440

64. Koyré (a) 237. Westfall 11 states that by 1661 only two significant figures had embraced the principle of inertia, Descartes and Huygens

65. Descartes (c) 85; (e) 38. See on Descartes' physics in particular: Scott

66. Descartes (c) 53, 65-73; cf. Kant (a) A20, 21, B5-6, 11-12, 36

67. Descartes (c) 74, 82; cf. Van der Hoeven 109-120

68. Descartes (c) 159; (e) 24-25

69. see Burtt 63-71, 83-90, 106-111, 115-121

70. Descartes (c) 53, 65-73

71. Galileo, *Il Saggiatore*, Drake (ed.) 273-278; cf. Koyré (a) 179

72. Galileo (b) 269-272 briefly discusses impact, announcing a separate treatise, the so-called Fifth or Sixth Day, published posthumously (1718)

73. Descartes (a) 21

74. Descartes (c) 86-88; Koyré (d) 77-78; Van der Hoeven 120-139

75. Descartes (c) 89-94. Descartes treated motion in terms of speed and not velocity. He was corrected by Huygens, see below
76. *ibid*. 93. See Hübner 304: "...Descartes' rules of impact describe fundamental processes within nature as God sees them."
77. cf. Koyré (d) 77; Harman 12
78. Galileo (a) 21-28; (b) 162-166
79. Descartes (c) 76-79
80. Kepler (b) 34 (Introduction), 228 (Ch.34); Galileo (a) 345
81. cf. Galileo, *Letters on the Sunspots*, Drake (ed.) 106; *Letter to the Grand Duchess Christina*, Drake (ed.) 212-213
82. Descartes (c) 210-214
83. Kepler (b) 25 (Introduction)
84. *ibid*. 26; cf. Koyré (c) 194
85. Hanson (c) 275 says that Kepler was no Copernican, because his three laws were not foreseen in Copernicus' *Revolutionibus*. But without any doubt Kepler considered himself a Copernican, and Newton calls Kepler's first and second laws: "The Copernican hypothesis", Newton (a) 395-396
86. Kepler (b) 21 (Introduction)
87. *ibid*. 186 (Ch.24), cf. M. Caspar's introduction, p.43*, and Koyré (c) 181
88. Descartes admitted that the planetary orbits are not perfectly circular, but did not accept Kepler's laws, see Descartes (c) 117
89. The Italian G.A.Borelli was the first astronomer to use Kepler's laws, in 1666. Newton used Borelli's results. Only about 1670 did scholars begin to consider Kepler's work seriously, among them Hooke, Halley, and Wren. Before Borelli, the Englishman J. Horrocks accepted Kepler's laws, but he died too early to exert any influence
90. Nevertheless, Newton read little or nothing of Kepler's works, see Cohen (e) 189. On the development of the concept of force, see Jammer (b) Ch.5-7; Cohen (d)
91. Descartes (a) 88
92. Kepler (b) 26 (Introduction); cf. Koyré (c) 194; see Kepler (b) Ch.32-39 on force; cf. Koyré (c) 185-224
93. Gilbert; Kepler (b) 229 (Ch.34), 329, 331 (Ch.57); Galileo (a) 399-414
94. Kepler (b) Ch.34, 57; cf. Koyré (c) 208
95. Newton (a) 409
96. Heilbron 19-43
97. Descartes (c) 278-305
98. Newton (a) 398-400. However, as a force, gravity is not a property of a body apart from other bodies by which it is attracted.

99. *ibid*.414
100.Drake (c) Ch.1
101.Descartes (c) 71-73; (e) 16-23
102.Pascal 233-259
103.Newton (a) 21
104. Cohen (b) 322-327; Harman 13-17; Dijksterhuis 512-515
105. For a discussion of Newton's laws of motion, see Hanson (b); Ellis; Nagel 174-202; Cohen (e) 171-193
106. Newton (a) 14-17
107. *ibid.* 21
108. *ibid.* 2-6. In fact it was Hooke who in 1674 first observed that circular motion requires an unbalanced force, see Westfall 382-383
109. Newton (a) 22
110. *ibid.* 13
111. *ibid.* 17
112. McMullin 2, 29-56
113. Stafleu (b). In contrast to Newton's dualism, we find the Cartesians' monism, stressing matter identified with space, and Leibniz' monism stressing force (dynamism); see Jammer (b) Ch.9
114. Szabo 47-85; Jammer (b) 165-166
115. Stafleu (d)
116. Stafleu (b)
117. Newton (a) 6-12
118. Kant (a) A19 ff, B33 ff recognized its relevance
119. Newton (a) 10-11
120. Alexander (ed.) The largest part of the debate concerned theological questions
121. According to Leibniz, space is "the order of co-existence", and time "the order of succession"
122. Mach 279-286; cf. Grünbaum (a) Ch.14
123. Mach 286-290
124. Galileo, *Il Saggiatore*, Drake (ed.) 277-278; cf. Koyré (a) 179
125. Dijksterhuis 503
126. Harman 18
127. Dijksterhuis 515

Chapter 4 : **The solution of problems**

1. Ravetz 72: "...science is a special sort of problem-solving activity...." Laudan 121 ff hypostatizes the solution of problems as the aim of science.
2. cf. Hanson (a) 99 ff for a discussion of the function of Newton's second law of motion in various contexts
3. Nagel 32

4. Bunge (c) 402

5. On the function of axioms in a theory, see Bunge (a), (b). In physics, axioms are called by many names: law, postulate, principle, rule, theorem, *etc.*

6. Bunge (a) 85 speaks of "protophysics" as the set of presuppositions of physical theories; Bunge (c) 402 defines a "lemma" as a theorem proved in an alternative theory

7. According to Sneed 16: "The essential, distinguishing feature of mathematical physics is that each has associated with it a formal mathematical structure, the core of the theory, or the mathematical formalism characteristic of the theory." This evidently exaggerates the importance of mathematics in physics

8. Ravetz 85 distinguishes between "data" and "information"

9. Bunge (c) 402

10. In Galileo (b) and Newton (a), a "proposition" is either a "problem" or a "theorem."

11. Koyré (d) 32: "All the meanings have this in common, that they attenuate (or suppress) provisionally (or definitively) the affirmative character and the relation to truth (or reality) of the "hypothetical" proposition. A hypothesis then is not a judgment, properly speaking, but an assumption or a conjecture that one examines in its consequences and implications, which should either "verify" or "falsify" it."

12. cf. Bunge (c) 226: "...in the logical sense of "hypothesis", all the initial assumptions (axioms) of a theory...are hypotheses."

13. Popper (b) Ch.10; Kuhn (b)

14. Popper (a) 19, 70, 241-244, 288-289, *etc.*

15. Popper (b) 312

16. *ibid.* 313. It should be observed that Popper, speaking of "theories," probably means "hypotheses"

17. Kuhn (b) Ch.2-4; (c)

18. Kuhn (b) 10, 23

19. cf. Masterman, who localizes more than twenty different meanings of "paradigm" in Kuhn's work. See also Suppe 135-151, 643-649; Toulmin (b) 98-130. Both stress that Kuhn's views have changed considerably in the course of time

20. Kuhn (b) 175

21. Kuhn could have added Tycho's, Galileo's and Descartes' works. Cf. Galileo (b) 242-243: "...in the principle (of accelerated motion) laid down in this treatise (Galileo) has established a new science dealing with a very old subject...he deduces from a single principle the proofs of so many theorems...the door is now opened, for the first time, to a new method fraught with numerous

and wonderful results which in future years will command the attention of other minds..."

22. Kuhn (b) 20, 187-191

23. *ibid* Ch.4. It is not quite clear to me whether Kuhn still maintains this view.

24. cf. Popper (c) and Watkins. Their criticism is justified if normal science would degenerate into the belief that alternatives are impossible, see Feyerabend (e) Ch.3; Lakatos (b) 68-69

25. Kuhn (b) 10

26. *ibid*. 37

27. Galileo (a) 261, 396

28. Laudan 17, 18, 26-30. For Kuhn, an anomaly is a problem which defies resolution in the context of an accepted paradigm.

29. Van Helden 150

30. Huygens 132-133; cf. Dijksterhuis 507-509

31. Bunge (c) 165

32. cf. Popper (b) 222; Laudan 108-109 speaks of the problem-solving capacity of a theory or "research tradition" as a criterion for the choice between theories or traditions. Also Lakatos pays attention to this, see Sec. 6.2. Kant (b) 352 observes there never comes an end to questions, cf. Rescher

33. Koestler 401

34. see axiom 4 in Copernicus' *Commentariolus*, Rosen (a) 58

35. Dijksterhuis 327

36. Copernicus, *Commentariolus*, Rosen (a) 58, 63. Only in his *Revolutionibus* did Copernicus react to the objection, see Copernicus 42-46 (I,7-9); cf. Koyré (a) 132-135

37. Copernicus 43-46 (I,8); cf. Koyré (a) 133

38. Galileo (a) First Day; Koyré (a) 135-141

39. Galileo, *Sidereus Nuncius*, Drake (ed.) 42-45; Galileo (a) 67-69, 91-99

40. Clavelin 199-203

41. Kepler (a) 129

42. Cohen (a) Ch.3-5; Westfall 402-404

43. Newton (a) 415-416

44. *ibid*. 478-484

45. Popper (d) 170

46. Popper (a) 269, (b) 36: "Confirmations should count only if they are the result of *risky predictions*; that is to say, if, unenlightened by the theory in question, we should have expected an event which was incompatible with the theory — an event which would have refuted the theory." See Lakatos (b) 38-39; Grünbaum (b)

47. Newton (a) 547
48. Popper (b) 63, 174; (d) 191; Hempel (c) 70
49. Nagel 42-46
50. Newton (a) 398-400
51. R. Cotes, Preface to the second edition of Newton (a) XX-XXXIII; Kant (a) B17-18, 21
52. Popper (a) 71-72; Bunge (c) 436-445
53. Braithwaite 2: "...the fundamental aim of science is the establishment of laws..."; Popper (d) 191: "...it is the aim of science to find *satisfactory explanations* of whatever strikes us as being in need of explanation", 193: "...(explanations)...in terms of testable and falsifiable universal laws and initial conditions"; Bunge (c) 27: "The primary target of scientific research is...the advancement of knowledge...scientific explanation and prediction are based on law statements, which in turn interlace in theories...factual science seeks to map the patterns (laws) of the various domains of fact"; Bunge (c) 345: "...the central goal of science is the *discovery* of objective patterns..."
54. Brown 166: "Our central theme has been that it is ongoing research, rather than established results, that constitutes the lifeblood of science. Science consists of a sequence of research projects structured by accepted presuppositions which determine what observations are to be made, how they are to be interpreted, what phenomena are problematic, and how these problems are to be dealt with."
55. Descartes (a) 3 asserts that his method leads to increasing knowledge, improving abilities, and progress
56. Burtt 36-38; Feyerabend (b), (d); Clavelin 58-60; Galileo (a) 56-57
57. Koyré (a) 136
58. Copernicus 51, 143 ff; Dreyer 330, 371; Koestler 204
59. Dreyer 345-360; Hall (b) 18-20
60. Kuhn (b) Ch.6-8
61. *ibid.* 68, 69
62. see, *e.g.*, the discussion in Beer and Strand (eds.), Session 3, in particular Gingerich; Rosen (b) 131-132
63. Dijksterhuis 325
64. Koyré (c) 94; Duhem (b) 70-74
65. Cf. Laudan 14ff, 45ff, 88, who distinguishes between "empirical problems" and "conceptual problems", only the latter giving rise to a crisis
66. Even the crisis leading to the disbandment of the Pythagorean brotherhood was caused by the theory leading to the Pythagorean

theorem, see Sec. 3.1

67. For an extensive discussion of revolutions in science, see Cohen (f).

Chapter 5: **The systematization of knowledge**

1. Koyré (d) Ch.1; Dijksterhuis 509-510; Cohen (e) 157-162 objects to the expression "Newtonian synthesis", "...because it tends to mask the creative way in which any scientist uses the work of his contemporaries and predecessors" (158)

2. Perhaps Newton himself aspired to achieve a synthesis of dynamics, gravity, optics, *and* alchemy, but he never succeeded in that

3. Duhem (a) 190-195; Popper (d) 197ff; (f) 139-144, 148. Hempel (b) 344 adopts a moderate view. See also Feyerabend (c) 168; (e) 35-36; Cohen (c); (e) Ch.5; Finocchiaro (a) 180-188, 196-198; Brown 60-66

4. Popper (d) 16, 198-200, 357; cf. Newton (a) 55

5. Newton (a) 21-22, 411-414 describes an independent experiment on pendulums, in order to confirm Galileo's law that the acceleration of gravity is independent of the mass of the falling body.

6. Newton was well aware of the fact that these empirical generalizations are approximations, see Newton (a) 405, 422. Cf. Glymour 222-224; Laudan 24. Also Galileo stressed the approximative character of his theory of projectile motion, see Galileo (b) 251

7. Hesse (b) 9-16. "Similarity" is the keyword in Hesse's network model of theories, see Hesse (b) Ch.2

8. Popper (a) 32-33 distinguishes four methods to test a theory: internal consistency, the logical form of the theory, comparison with other theories, empirical application of its conclusions

9. Popper (a) 86ff states that theories can be refuted by *accepted* basic statements. This implies a subjective element, cf. Popper (a) 108ff

10. Newton (a) 401-405 (Kepler), 411-414 (Galileo)

11. This method corresponds with Hempel's "hypothetical-deductive method", see Hempel (b) 365; (c) 9

12. On the distinction between "naive" and "sophisticated", "dogmatic" and "methodological" falsificationism, see Lakatos (b) Ch.1; Brown 76; Bunge (c) 266-269; (d) 324

13. Reichenbach (a) 87-88

14. Bunge (d) 323

15. Popper (a) 50

16. Grünbaum (b)

17. Popper (a) 82

18. *ibid.* 86-87

19. Brown 71ff; Bunge (d) 316

20. Popper (b) 102; Galileo (a) 327-328, 339; p. 328: Copernicus *et al.* "...have through sheer force of intellect done such violence to their own senses as to prefer what reason told them over that which sensible experience plainly showed them to the contrary."

21. Copernicus 49, 51 (I,10); Galileo (a) 372-389

22. Burtt 38

23. Lakatos (b) 16-17, 40-41; Popper (a) 50; Polanyi 20

24. Lakatos (b) 35

25. Descartes (a) 1,77

26. Koyré (a) 4

27. Plato 365-370 (*Meno*); Galileo (a) 169-180

28. Galileo (a) 12, 22, 89-90, 145, 158, 191, 291

29. Galileo (b) 42

30. Hempel (c) 15, and Popper's method of Conjectures and Refutations stress the importance of imagination

31. Plato 762 (*Republic* VII)

32. Descartes (c) 14, 26

33. Bunge (d) 162

34. Galileo (a) 336, and note to p.320; Drake (ed.) 73. Of course, Galileo pointed out that besides Jupiter no other planet has four moons, and that these moons are in motion. It is not easy to blame that on the telescope.

35. Kuhn (b) 16-17

36. On measurement, see Stafleu (a) Ch.3

37. Galileo (a) 315; Westfall 540-548, 583-586

38. Galileo (a) 289-290

39. On experiments, see Kuhn (d) Ch.3,8

40. Galileo (b) 251-257 discusses the disturbing factors in experiments on ballistic motion

41. Hooykaas (b) Ch.4

42. Dijksterhuis 153. Even Leonardo da Vinci did not distinguish between experience and experiment

43. The most important works by Francis Bacon are *Advancement of Learning* (1605) and *Novum Organum* (1620). His unfinished work intended to bring about an *"Instauratio Magna"* (Great restoration); cf. Blake *et al.*, Ch.3

44. Galileo (a) 408, 410; Dijksterhuis 374; Shea 37-44, 92; Koyré (c) 165-173, 469; Drake (c) 44

45. Galileo (a) 62, 144-149, 256

46. Galileo (b) 276

47. Descartes (a) 63, 65; Hall (a) 184; Blake *et al.*, Ch.4

48. Suppe 45-49, 80-86
49. Lakatos (b) 44; Hanson (a) Ch.1
50. cf. Popper (b) 112, 238-240
51. Galileo (b) 264 stresses the need of a common standard for velocities, just as for intervals of time
52. Popper (a) 44ff; Kant (a) A820, B848

Chapter 6: **Heuristics**
1. In a narrow sense, "heuristic" is the art of solving problems
2. On Aristotle's method of science, see Randall 52-56
3. Descartes (a) 21, 29; (c) 16
4. see Popper (a) 27-30; (d) Ch.1; (f) 11-158. For a critical discussion see Grünbaum (c)
5. cf. Bunge (c) 314-323; (d) 290-294; Ziman (c) 43-56
6. The problem of "curve fitting" never found a satisfying logical justification, according to Glymour 322-340; cf. Ravetz 84. On "inductive logic", see Finocchiaro (b) 293-297
7. Popper (a) 30-31
8. Popper (b) 187-188; (d) 173
9. Kepler (b) 5-12 (Dedication); cf. Koyré (c) 277-278
10. Hanson (a) 72ff; Simon 41-43
11. see Kepler's footnotes (1621) to Kepler (a) (1597)
12. cf. Cohen (c) 312-313; Westfall 446-452
13. Popper (a) 50; cf. Brown 72
14. Popper (f) 118, 122
15. cf. Bunge (c) 244; (d) 322
16. Koestler 11, 333ff, 400
17. Reichenbach (a) 6-7; (b) 231; Popper (a) 31; Urbach
18. Kuhn (b) 10-11; Hanson (a); Feyerabend (e); Lakatos (b) Ch.1; Laudan 78ff
19. Popper, in: Schilpp (ed.) 999
20. Lakatos (b); (c)
21. Lakatos (b) 68-73, 117; (c) 110; see Musgrave; Feyerabend (f)
22. Lakatos (b) 33-35. On a proposal by Zahar, Lakatos later defined a "novel fact" as a fact not used in the construction of a theory, see Lakatos (b) 184-185; Zahar; Worrall
23. Lakatos (b) 33-34
24. see, *e.g.*, Howson (ed.)
25. Lakatos (b) 50, 51; cf. Cohen (e) 62-68
26. Lakatos (b) 51
27. see, *e.g.*, Lakatos (b) 55-68 on Bohr's research programme
28. Bohr's "Principle of Correspondence", mentioned in Sec. 5.1, concerns a particular example of the method of successive approxi-

mation
29. Lakatos (b) 168-192
30. *ibid.* 188. Copernicus' so-called trepidation theory, see Sec. 4.5, intended to explain a non-existent effect with Ptolemy's methods, earned him the admiration of several 16th-century astronomers, who, therefore, did not consider this stage to be "degenerative" in Lakatos' sense
31. Koyré (d) 30, 102; Cohen (e) 229; Westfall 444-445
32. Newton (a) 395-396 calls Kepler's first and second law: "The Copernican Hypothesis", see Koyré (d) 101-103. Apparently, Newton considered it an empirical generalization, in this context useful to refute the Cartesian vortex theory
33. Kepler's third law was less suspect than the first and second, and was easier to establish and to represent in a table, convincing even people outside the limited circle of professional astronomers, see Newton (a) 401-405; cf. Cohen (c); (e) Ch.5; Glymour 207-209
34. Newton (a) 419-420
35. Bunge (c) 18 observes that the method of successive approximation is characteristic, though not exclusively, of science. Suppe 670: "...it is at best *a* pattern of good reasoning, and very often science does, and ought to, employ other, incompatible, patterns in proceeding rationally."
36. cf. Dooyeweerd, in particular vol. II
37. Galileo (a) 115-116, 126-127, 155, *etc.* Before Galileo, Bruno used the analogy of a moving ship to refute the arguments against the earth's motion, see Koyré (a) 136-138
38. Galileo (a) 126, 141-145, 148, 154-155, *etc.*
39. *ibid.* 424-425
40. Koyré (a) 247
41. Koyré (d) 7
42. Galileo (a) 12-14
43. Galileo (b) 90-91, 137; cf. Koyré (a) 201; Cohen (e) 17-38, Ch.3. The 14th-century scholar Oresme was the first to use graphs in the derivation of mechanical laws. Galileo knew his methods. Besides Oresme also the so-called *Calculatores* from 14th-century Oxford were concerned with the mathematization of mechanics, see Dijksterhuis 207-220
44. Glymour 203-226
45. Newton (a) 398-400
46. *ibid.* 40-45 observes that this "fixed point" may move uniformly and even with acceleration, if there is an external force. Newton needs this in order to apply his theory to the system of earth and moon, moving together around the sun.

47. *ibid.* 46 Newton found the formula for centripetal acceleration independently of Huygens.

48. *ibid.* 303-304, 428-433

49. Hesse (a) 146 interprets "weight" as "gravitational mass", which contradicts Newton's statement (Newton (a) 303) that weight equals motive force, and (Newton (a) 428-433) that the weight of an object depends on its place on earth. Cohen (e) 271-273 makes a similar mistake.

50. Bunge (c) 400. In his popular expositions of relativity theory, Einstein ascribed the distinction between inertial and gravitational mass to "classical mechanics." In fact, both in classical (Newtonian) mechanics and in the general theory of relativity, it is an *axiom* that all bodies are equally accelerated in a gravitational field. But Einstein drew different conclusions from this axiom than Newton did.

51. Popper (b) 106

52. cf. Galileo (b) 1

53. Ziman (b); Pacey; Price

54. cf. Kuhn (b) 26

55. Koyré (c) 178

56. Kepler (b) 166 (ch.19), cf. 264 (Ch.43). Translation: Koestler 326-327, cf. Koyré (c) 178. The translation is far from literal

57. Drake (b) 86-90; cf. Galileo (a) 387 on the accuracy of instruments

58. Newton (a) 411

59. *ibid.* 430

60. Feyerabend (e) Ch.1; (g). For a criticism of Feyerabend's views on Galileo, see Finocchiaro (b) 182-200

61. Laudan 95-100, 103-105

62. cf. Stafleu (b); (c); (d)

63. Stafleu (a) Ch.1

64. *ibid.* 7

Chapter 7: **The principle of clarity**

1. Popper (b) 313

2. *e.g.,* Descartes (a) 57-59

3. Suppe 16, 45ff, 66ff; Von Mises; Bunge (b). It is doubtful whether any physical theory can ever be completely rendered in formalized language, see Stegmüller 3-7

4. Suppe 102-109; Braithwaite Ch.3 rejects the possibility of defining theoretical terms in the observation language: "A scientific theory which, like all good scientific theories, is capable of growth, must be more than an alternative way of describing the generalizations upon which it is based, which is all it would be if its theoretical terms were limited by being explicitly defined." (76)

5. Bunge (c) 47: science has a language, but it is not a language, it is a body of ideas and procedures expressed in a number of languages

6. Ziman (a) Ch.6; (b) Ch.5; (c) Ch.2

7. Copernicus 27 (Preface)

8. Newton (a) 397

9. Drake (c) 51

10. Descartes (a) 77-78

11. Descartes' *Principia Philosophiae* (1644) was translated into French in 1647. Already in 1377 Oresme wrote his *Traité du Ciel et du Monde* in French. It contains a translation of Aristotle's *On the Heavens*. Paracelsus also wrote some of his medical works in his native language, "Swabian" German

12. cf. Koyré (c) 514: "Most people in the seventeenth century wrote atrociously badly. Writers of the standard of Galileo and Torricelli were quite rare exceptions."

13. Galileo, *Il Saggiatore*, Drake (ed.) 237-238

14. Galileo (a) 345-356. See footnote 67 to Chapter 2.

15. Finocchiaro (b) 311

16. Galileo (a) 401-403

17. cf. Macey Ch.4, 5

18. see Koyré (c) 437-438

19. cf. Galileo, *Il Saggiatore*, Drake (ed.) 270-273

20. Drake (b) 169-205; (c) Ch.8

21. Drake (ed.) 84

22. Heilbron 13: The chief agent in changing the scope of physics was the demonstration experiment

23. Copernicus 50 (I,10)

24. Also Cotes' Preface to Newton (a), second edition, is full of polemics

25. Newton (a) 543

26. *ibid.* 399

27. Galileo considered himself the discoverer of all novelties in the sky, including the sunspots, see Galileo (a) 345

28. Newton (a) 491-498; cf. Galileo (a) 52; Shea 86-87, 105

29. Grassi replied in 1626 with *Ratio Ponderum Librae et Simbellae*, after he had become friends with Guiducci in 1624

30. Finocchiaro (b) 275

31. Simplicius was the last ancient philosopher. He defended the division of labour between astronomy and physics (see Sec. 2.1), cf. Dreyer, 131-132, Duhem (b) 5-11, 23-24. Salviati and Sagredo

were friends of Galileo, both deceased before 1632

32. Finocchiaro (b) 12-16. Salviati maintains to be impartial, see Galileo (a) 107, 131, 256, 274, 356, 369, 413, 463

33. Galileo (a) 277; Finocchiaro (b) 47

34. Galileo (b) 205

35. Koyré (a) 188; Drake (ed.) 81

36. Galileo (a) 326 rejects Tycho's system without mentioning his name, after Simplicio had discussed it, p. 322-326

37. For this reason, Finocchiaro (b) 44, 167 defends Galileo's omission of Tycho's system

38. Finocchiaro (b) 22

39. Galileo (a) 390-399 on the Copernican system, 342-345 on retrograde motion, cf. Copernicus 291-294 (V,35); see Koyré (a) 158-159

Chapter 8: **Science and society**

1. Drake (c) Ch.4; see Drake (ed.) 77-78 for the constitution of the *Accademia dei Lincei*

2. Ziman (b)

3. cf. Agassi, who also distinguishes between "inductivist" and "conventionalist" historiographers; Basalla; Koyré (d) 5,6; Finocchiaro (a) Ch.7; Kuhn (d) Ch.5; Hooykaas (c); Lakatos (b) 118-121; Laudan Ch.8

4. Hooykaas (c) 178-209; Büchel 66-70; Basalla

5. Basalla; Hooykaas (b) Ch.5; Koyré (c) 74-75

6. Zilsel; Büchel 51-86; Drake (c) Ch.1; Wightman

7. Gimpel; Grant; Lindberg; Pacey; Price; Macey

8. Galileo, *Letter to the Grand Duchess Christina*, Drake (ed.) 182, citing Tertullian: "We conclude that God is known first through Nature and then again, more particularly, by doctrine; by Nature in His works, and by doctrine in His revealed work."

9. Hooykaas (b) 121-122; Blumenberg 371-395

10. Kuhn (a) 191-192 quotes A.D. White's *A History of the Warfare of Science with Theology in Christendom* (1896), whose unreliability has been demonstrated by Hooykaas (b) 121

11. Galileo, *Letter to the Grand Duchess Christina*, Drake (ed.) 186; cf. Kepler (b) 29-33; Drake (ed.) 169, 181-184

12. Council of Trent: 1545-1563; Jesuit order founded: 1540

13. Drake (ed.) 1-58

14. *ibid.* 162-164

15. *ibid.* 143-216

16. A much better interpretation of Joshua 10 was given by Kepler (b) 29-30

17. Drake (b) 252-256

18. Galileo (a) 357-358
19. *ibid.* 462
20. Drake (b) 341-352; de Santillana. For the (partial) text of the indictment, see note to page 103 of Galileo (a), and for the text of Galileo's abjuration, see Galileo (a) xxiv
21. In 1635, a clandestine Latin translation of the *Dialogue* was published, which was corrected by Galileo in 1641
22. The *Discorsi* was announced in the *Dialogue*, Galileo (a) 452, 464-465
23. *e.g.*, the Italian astronomer Borelli, see Koyré (c) 471
24. Pascal 467 (my translation)
25. Ziman (a)
26. cf. Drake's introduction to Galileo (a) xxiv
27. Galileo (b) xix
28. Descartes (a) 22, 23
29. It might be suggested that the "scientific church" has already convicted its own Galileo in the person of Velikowski, see Brown 160-165

Chapter 9: **Parsimony and harmony**
1. Mach 577-595
2. *ibid.* 577
3. *ibid.* 586
4. *ibid.* 578-579
5. *ibid.* 579
6. *ibid.* 582
7. *ibid.* 588
8. *ibid.* 586-587
9. *ibid.* 583
10. *ibid.* 588-590
11. *ibid.* 284
12. Copernicus, *Commentariolus*, Rosen (a) 30. Descartes (c) 109 states that Copernicus' system is more simple and clear than Tycho Brahe's
13. Burtt 38; Butterfield 29. The numbers "34" and "80" are by no means invariant in the literature. Margenau 97 bears the palm with: "Copernicus, by placing the sun at the center of the planetary universe, was able to reduce the number of epicycles from 83 to 17".
14. Koestler 194-195; Gingerich arrives at a slightly lower number; cf. Neugebauer (a) 204
15. Dijksterhuis 325; Kuhn (a) 168,171; Feyerabend (b); Gillispie 24-26; Toulmin, Goodfield 175, 179; Koestler 194-195, 579-580; Koy-

ré (c) 43. It should be added that to ascribe the *daily* motion to the earth alone instead of to *all* celestial bodies makes Copernicus' system kinematically much less complex than that of Ptolemy or Aristotle.

16. Koyré (c) 43, 45
17. Galileo (a) 342-344
18. see Bunge (d) 347-349
19. Feyerabend (b) says that the realists were in favour of Tycho. Only instrumentalists could be in favour of Copernicus.
20. Galileo (a) 464; see Finocchiaro (b) 8-12
21. cf. Lakatos (b) 184 ff
22. Popper (b) 97-100; see also Popper (f) 111-131 for a criticism of instrumentalism
23. Blake *et al.* 31-35
24. Newton (a) 410, 414-415, 418-419
25. Kepler (a) 31
26. Kepler (b) 23 (Introduction)
27. Galileo (a) 397
28. *ibid.* 418
29. Galileo (b) 161
30. Galileo (a) 347-355; cf. Drake (c) 191-196
31. Newton (a) 398
32. Copernicus 50 (I,10)
33. Rheticus, *Narratio Prima*, Rosen (a) 164-165
34. Kepler (c) 289 (V,3)
35. *ibid.* 291
36. *ibid.* 279-280 (Preface to book V), translation quoted from Koestler 399; cf. Koyré (c) 343, 457
37. *e.g.*, Galileo (a) 118-119
38. *ibid.* 31
39. Clavelin 215
40. Galileo (a) 259
41. Galileo (a) 28; (b) 215, 244, 251
42. Galileo, *Letters on the Sunspots*, Drake (ed.) 113-114; Galileo (a) 31-32, 147; cf. Clavelin 372-374
43. Galileo (b) 192-193; Koyré (c) 119. However, Drake (c) Ch.12, 13 argues against the view that Galileo was obsessed by uniform circular motion, that he adhered to circular inertial motion, and that he seriously believed the planets to move in uniform circular motion around the sun
44. Galileo (b) 193-194
45. Cohen (e) 20-21; Van Helden
46. Drake (c) 53

47. Galileo (a) 133

Chapter 10: **Criticism**
1. Popper (b) 316
2. Descartes (c) 48, 53-54
3. Descartes (d) 351-355, 359-362. Popper (d) 155-156
4. Popper (f) xiv-xxv, 159-193; cf. Mach 586-587
5. Popper (f) 194-216
6. Popper (a) 37-38
7. Popper (f) 181. In particular in mathematics, an existential statement can sometimes be refuted by a *reductio ad absurdum* argument.
8. *ibid.* 185-186
9. *ibid.* 178-179
10. *ibid.* 198-199
11. cf. Lakatos (a) 6-7
12. Kant (a) B14
13. Popper (f) 256-258, 235
14. Hempel (b) 1-51; Brown 25-32
15. Bunge (d) 315-317
16. Popper (f) 29; for Kant, transcendent criticism meant criticism based on non-scientific arguments.
17. Newton (a) 325
18. *ibid.* 392-396
19. *ibid.* book II
20. *ibid.* 396
21. *ibid.* 543
22. Popper (f) xxvii,xxxi,361
23. Popper (a) 104, 108; Lakatos (b) 22-28, 37-47
24. Ziman (a) Ch.6
25. Copernicus 26-27 (Preface)
26. *ibid.* 27
27. Koestler 150-157
28. Newton (c) Ch. II
29. Koestler 381-383

Chapter 11: **Commitment**
1. On charity in reasoning, see Finocchiaro (b) 340, 425
2. On "hypotheses" during the Copernican revolution, see Blumenberg 341-370
3. Copernicus 22-23; see Koyré (c) 36-37; Duhem (b) Ch.6
4. Copernicus 36. This passage only occurs in the manuscript, not in the printed work

5. Copernicus, *Commentariolus*, Rosen (a) 22-33; cf. Koyré (c) 85-86; (d) 31-32

6. Kepler (b) 4. Shortly after the printing of *Revolutionibus*, Rheticus protested against the anonymous Preface, but in vain. See Rosen (b) 85,125-128, 167-168, 192-205

7. Kepler (b) Introduction. For Galileo on Osiander's preface, see Drake (ed.) 167-168. Kepler on hypotheses, see Kepler (a) 30-31, 39; (b) 171-175 (Ch.21); cf. Hooykaas (c) 39-59; Blake *et al.* 37-43; Koyré (c) 95-97; Burtt 65. Duhem (b) 100-104. For a translation and a discussion of Kepler's *Apologia*, see Jardine.

8. Popper (b) 98, Blake *et al.* 28, Toulmin (a) 39-43 all discuss Osiander's instrumentalism, each without mentioning Osiander's main argument, cf. Feyerabend (d) 327

9. Galileo, *Letters on the Sunspots* (1613), Drake (ed.) 93-94; Galileo (a) 322, 324

10. According to Dreyer 325, Plato already assumed that the planets are not primary light sources

11. Galileo (a) 339; Galileo (a) 325, 334-339 proves that the variations in the brightness of Venus and Mars constitute arguments in favour of Copernicanism. Descartes (c) 108-109 calls the phases of Venus the main argument against the Ptolemaic system

12. Galileo (a) 5, 6

13. Finocchiaro (b) 17

14. *ibid.* 212-213

15. On "rational reconstructions" see Lakatos (b) Ch.2

16. This was once more stressed by Regiomontanus; see Dijksterhuis 322, also for our "rational reconstruction"; Koyré (c) 103-104

17. Copernicus, *Commentariolus*, Rosen (a) 58-59, mentions seven "assumptions which are called axioms."

18. Blumenberg 281

19. Descartes I 271

20. Descartes (e) published posthumously in 1664. Parts of *Le Monde* can be found in Descartes' *Principes de la Philosophie*

21. Descartes (c) 109-110, 113, 115-116

22. see, *e.g.*, Sabra; Ronchi; Whittaker; Scott Ch.4

23. Kepler's *Ad Vitellionum paralipomena, quibus Astronomiae pars Optica traditur* (1604) gives the modern explanation of the physical functioning of the human eye. His *Dioptrice* (1611) contains the first theory of the telescope and the invention of the so-called astronomical telescope

24. Aristotle (c) II,7

25. Descartes (a) 42, 81

26. Descartes I 307; (a) 43, 84; (c) 136; (e) 98; cf. Duhem (a) 33-34

27. Descartes (a) 93-105. Descartes' proof remained a matter of dispute, and at least three alternatives were proposed. All agreed that the index of refraction be equal to the ratio of the speed of light in the two media concerned. For the refraction of light from air to water, Descartes and Newton took it to be the speed in water divided by the speed in air, but Fermat and Huygens found the inverse ratio. Only in 1850 could Foucault experimentally decide in favour of the latter opinion.

28. Sabra 469; Van der Hoeven 232-236

29. Descartes (c) 123-126

30. Galileo (b) 42-44

31. Newton (b) 1, cf. 388

32. Cohen (a); Burtt 215-220; Koyré (d) Ch.2, 6; Blake *et al.*, Ch.6

33. see Newton (c) 47-238 for his optical papers

34. As a mechanical philosopher, Newton did not really believe light to be coloured, but to be perceived coloured, colour being a secondary property

35. Newton (a) 385, 395, 419, 489; cf. Koyré (d) 29. In the first edition, Newton called his axioms still "hypotheses"

36. Newton (a) 547

37. *ibid.* The reference to "electric bodies" is only found in Motte's translation of *Principia*

38. Newton (b) 339. For the Queries, see p. 339-406

39. Query 31 takes about 30 pages, Newton (b) 375-406

40. *ibid.* 404

41. Newton (a) 547

42. For a review of Aristotle's, Descartes' and Newton's opinions on gravity, see Heilbron 19-63

43. Galileo (a) 20; (b) 87

44. Newton (a) 399-400

45. Newton also required positive evidence for hypotheses, rejecting the method of *reductio ad absurdum* demonstrating the truth of a statement by proving the absurdity of its alternatives

Chapter 12: **Belief**

1. Hooykaas (a) 74-75; Knowles; Grant

2. see in particular Plato's *Republic* V-VII; cf. Russell Ch.15

3. Galileo's Platonism is disputed by Clavelin 424-431, Finocchiaro (b) 159 and Drake (b) xxi, but affirmed by Burtt, Ch.3, and Koyré (a) 159, 202-209, 223

4. cf. Galileo (a) 12, 22, 89-90, 145, 158, 191-192

5. *ibid.* 256; cf. Feyerabend (d)

6. Galileo (a) 103

7. Aristotle (b) II,1

8. *ibid.* III,1

9. Galileo, *Letter to the Grand Duchess Christina*, Drake (ed.) 169

10. Popper (b) 223-228; (d) 44-47, Ch.9

11. Laudan 121ff

12. Hacking

13. Reichenbach (a); cf. Braithwaite; Hesse (b)

14. Popper (a) Ch.6-8. Popper also rejects the "essentialism" of Aristotle and Descartes, see Popper (b) 103-107; (d) 194-195; (f) 1-30

15. Popper (f) 254-255

16. Galileo, *Letter to the Grand Duchess Christina*, Drake (ed.) 182

17. Descartes (a) 41-45; (c) 83-86; (e) 37

18. Copernicus 25 (Preface), cf. 51 (I,10): The order of the heavenly spheres: "Such truly is the size of this structure of the Almighty's."

19. Bunge (c) 345

20. Popper (d) Ch.5; (f) 131-149

21. Laudan 81-82, 86-93; cf. Brown 166

22. Dijksterhuis 503

23. *ibid.* 548-549

24. Kepler (a) 60-62; (c) 362; cf. Galileo (a) 110; Descartes (a) 9

25. Westfall 301

26. Macey, Ch.4

27. Descartes (a) 54: The rules of mechanics are the same as the rules of nature. Descartes (c) 37 rejects the search for final causes

28. *e.g.*, Kuhn (a) 3; cf. Burtt, 18-20

29. Koyré (c) 114-115; Lovejoy 101-108

30. Galileo (a) 37

31. On Descartes' and Newton's concepts of time and space, see Koyré (b); (d) 79-95; Capek; Jammer (a); Burtt Ch.4,7

32. Galileo (a) 319-320 observes that there is no proof that the universe is finite. Aristotle's assumption that the universe is finite and has a centre depends on his view that the starry sphere moves

33. Aristotle (b) IV,2,4

34. Also Galileo was aware of the principle of a "Cartesian" coordinate system, see Galileo (a) 12-14

35. Newton (a) 6-12; cf. Grünbaum (a) Ch.1

36. Kant (a) A 24, 34, B 38-39, 50

37. Koyré (d) 14

38. Galileo (a) 71-83

39. Descartes (c) 75: All varieties of matter depend on the motion of its parts

40. Randall 143

41. *ibid* 144: nous nousing nous, *i.e.,* thought thinking about thought; Aristotle (a) XII,7

42. Descartes (a) 23; (c) 325

43. Descartes (c) 88-90. According to Leibniz our world is the best conceivable

44. cf. Descartes (c) 26

45. I doubt (or: I think), hence I am. Descartes (a) 31-33; (b) 13-18; (c) 28-29

46. Descartes (a) 33-40; (c) 33-35

47. Descartes (c) 37-38

48. *ibid.* 31; Descartes (e) 36-37

49. Descartes (a) 1-2, 38-40

50. *ibid.* 2, cf. 46, 56-58, where Descartes compares the body with a machine, and states that it is the use of language which makes men different from machines. For Descartes, using language means having reason

51. Descartes (c) 29

52. Descartes (a) 21, 33, 38, 41; (c) 38, 43

53. Descartes (c) 44

54. Descartes (e) 36-38 derives the law of conservation of motion from the immutability of God

55. Kant (a) Introduction A11, B24, 25

56. Popper (f) 87, 316, 339

57. *ibid.* 334, 339; cf. Kant (a) A11, B25

58. Popper (f) 20, 23

59. *ibid.* 26

60. *ibid.* 71-158

61. *ibid.* 74, 128

62. *ibid.* 149

63. *ibid.* 150

64. *ibid.* 150-153

65. Dooyeweerd vol. I, part I

66. Popper (f) 27, 154

67. *ibid.* 157, 259-261

68. Dooyeweerd vol. II 36-49, 565-582

Index/Bibliography

Achilles — 60,224

Adler, A. (1870-1937) — 42

Agassi, J.L., 'Towards an Historiography of Science,' *History and Theory*, Beiheft 2 (1963) — 280

Albert of Saxony (*c*.1316-1390) — 268

Alembert, Jean-Baptiste le Rond d' (1717-1783) — 79

Alexander, H.G. (ed), *The Leibniz-Clarke Correspondence*, Manchester 1956 (1717) — 270

Alfonso X of Castille (1252-1284) — 105,206

Andersson, G. — see Radnitzky

Archimedes of Syracuse (287-212 BC) — 23,24,54,72,74,75,164,233

Aristarchus of Samos (*c*.310-230 BC) — 18,19

Aristotle of Stagira (384-322 BC)

    (a) *Metaphysics* (R. Hope, transl.), Michigan 1968 (1952) — 262,266,287

    (b) *Physics* (P.H. Wickstead, F.M. Cornford, transl.), books I-IV, London 1970 (1929), books V-VIII, London 1980 (1934) — 54,58, 90,234,266,286

    (c) *On the Heavens* (W.K.C. Guthrie, transl.), London 1971 (1939) — 34,35,54,264,266,267,279,284

    actual (potential) — 58,241,266

    astrology — 55,215

    atomism — 60,246

    cause (motion, process) — 55,57-59,61,63,67

    change — 21,56-60,62,68,234,236,249

    common sense (intuition) — 11,58,60,77,98,100,101,118,207,236, 239,247,249

    cosmology (world view) — 27,35,37,38,46,54,55,61,62,73,103-107,113,166-169,178,190,191,207,219,220,224,235,244,245, 247,248, 250,258

    elements — 54,59,60,163,224

    empiricism (observation) — 12,57,74,117,119-121,234,235

    essence (essential, accidental) — 21,22

form (matter, substance) — 31,57,58,68,79,119,233,234,239,243, 250,266

gravity (levity) — 22-24,44,54,55,64,65,118,164,229,285

homocentric (crystalline) spheres — 27,32-34,92,191,196,206, 219,220,262,282,286

light — 60,224

local motion— 21,55,56,58,62

logic (axioms) — 24,36,59,98-103,118,129,151,226,227,236,248, 249

Lyceum — 171

mathematics (quantity, measurement) — 22,37,55,122,143,168, 199,236

natural (motion, place) — 21,54-56,59,65,68,77,248,266

occult (manifest) qualities — 68,74

parallax — 264

prime mover — 55

projectile motion— 55,56,60,246,266

realism (philosophy) — 10,11,29,32,34,57,90,92,103-106,138, 164,168,170,172-176,178,219,220,233-235,241,243,247

relations — 22

space (place) — 54,244,245,248

theology — 35,55,249,250

time — 81

vacuum (void) — 75,236,245,246

violent motion (force) — 54-56,73,74,77,207,248

Averroës (Ibn Rushd) (1126-1198) — 34,219,220,233

Bacon, Francis (1561-1626) — 10,107,123-125,129,155,268,275

Bacon, Roger (*c.*1210-*c.*1292) — 123

Bär (Ursus), Reymers (1550-1599) — 217,218

Barberini, Francesco (1597-1679) — 181

Barberini, Maffeo, Pope Urban VIII, 1623 (1568-1644) — 168,169, 180,181,192

Basalla, G. (ed.), *The Rise of Modern Science*, Lexington, Mass. 1968 — 280

Beeckman, Isaac (1588-1637) — 10,67,100,104,176,213,224,225,242, 268

Beer, A., Beer, P. (eds.), *Kepler*, Oxford 1975

Beer, A., Strand, K.A. (eds.), *Copernicus*, Oxford 1975 — 273

Beer, P. — see Beer, A.

Bellarmine, Robert (1542-1621) — 51,176,179,180,192,219

Benedetti, Giovanni Battista (1530-1590) — 23,24,64,65,74,104,233

Bentley, Richard (1662-1742) — 160

Bernardini, G. — see Fermi
Bernoulli, Daniël (1700-1782) — 137,148
Bernoulli, Johannes (1667-1748) — 137,148
Bessel, Friedrich Wilhelm (1782-1846) — 46,95
Blake, R.M., Ducasse, C.J., Madden, E.H., *Theories of Scientific Method*, Seattle 1966 (1960) — 275,282,284,285
Blumenberg, H., *Die Genesis der kopernikanischen Welt*, Frankfurt a.M. 1975 — 265,280,283,284
Bohr, Niels (1885-1962) — 112,174,276
Bos, H.J.M., Rudwick, M.J.S., Snelders, H.A.M., Visser, R.P.W., *Studies on Christiaan Huygens*, Lisse 1980
Borelli, Giovanni Alfonso (1608-1679) — 150,269,281
Boyle, Robert (1627-1691) — 10,109,172,173,224,235,236,240,246, 248
Brahe, Tycho (1546-1601) — 9,49,71,172,176,177,192,197,217,248, 265,271
    astrology — 215,243
    comet (nova) — 62,87,107,166,167,236
    measurement (observation) — 39,47,50,105,114,122,126,131, 132,148-150,156
    observatory — 148,172
    parallax — 47,116
    system — 27,39,46,47,61,90,113,126,133,167,169,176,182,190, 207,217,218,220-222,264,280-282
    trepidation — 104
Braithwaite, R.B., *Scientific Explanation*, Cambridge 1968 (1953) — 261,273,278,279,286
Bridgman, Percy W. (1882-1961) — 23
    The Logic of Modern Physics, New York 1954 (1927) — 262
Broad, C.D., *Leibniz*, Cambridge 1975 — 262
Brody, B.A. (ed.), *Readings in the Philosophy of Science*, Englewood Cliffs, N.J. 1970
Brouwer, L.E.J. (1881-1966) — 261
Brown, H.I., *Perception, Theory and Commitment*, Chicago 1977 — 261,263,273-276,281,283,286
Bruno, Giordano (1548-1600) — 96,103,104,177,277
Büchel, W., *Gesellschaftliche Bedingungen der Naturwissenschaft*, München 1975 — 280
Bunge, Mario — 240
    (a) *Foundations of Physics*, Berlin 1967 — 261,262,271
    (b) 'Physical Axiomatics,' *Rev. Mod. Phys.* 39 (1967) 463-474 — 262,271,278
    (c) *Scientific Research*, vol. I, *The Search for System*, Berlin

1967 — 261-263,271-274,276-279,286
(d) *Scientific Research*, vol. II, *The Search for Truth*, Berlin
1967 — 264,274,275,282,283
(ed.), *The Critical Approach to Science and Philosophy*,
London 1964
Buridan, Jean (*c.*1295-*c.*1358) — 35,36,56,198,263
Burtt, E.A., *The Metaphysical Foundations of Modern Physical
Science*, Garden City, N.Y. 1954 (1924) — 265,273,275,281,284-
286
Butterfield, H., *The Origins of Modern Science, 1300-1800*, Toronto
1968 (1949) — 281

Calvin, Jean (1509-1564) — 177,178,180,239,256,257
Campani, Giuseppe (1635-1715) — 150
Cantore, E., *Scientific Man*, New York 1977 — 262
Capek, M. (ed.), *Concepts of Space and Time*, Dordrecht 1975 —
286
Carnap, R., *Philosophical Foundations of Physics*, New York 1966
— 264
Caspar, M. — 269
Cassini, Giovanni Domenico (1625-1712) — 40,150,172,173
Cesi, Federico (1585-1630) — 173
Christina, Grand Duchess of Tuscany (end 16th century-1637) —
179,267,269,280,286
Clarke, Samuel (1675-1729) — 82,160,161
Clavelin, M., *The Natural Philosophy of Galileo*, Cambridge,
Mass. 1974 (1968) — 263,266-268,272,273,282,285
Clavius, Christophorus (Klau) (1537-1612) — 176,192
Cohen, I.B. (a) *Introduction to Newton's Principia*, Cambridge
1971 — 285
(b) 'History and the Philosophy of Science,' in: Suppe (ed.)
308-349 — 262,270
(c) 'Newton's Theory vs. Kepler's Theory and Galileo's
Theory,' in: Elkana (ed.) 299-338 — 274,276,277
(d) 'Kepler's Century: Prelude to Newton's,' in: Beer and Beer
(eds.) 3-36 — 269
(e) *The Newtonian Revolution*, Cambridge 1980 —
261,269,270,274,276-278,282
(f) *Revolution in Science*, Cambridge, Mass. 1985 — 261,274
Cohen, M.R., Nagel, E., *An Introduction to Logic and Scientific
Method*, London 1978 (1934) — 261,262
Cohen, R.S., Feyerabend, P.K., Wartofsky, M.W. (eds.), *Essays in
Memory of Imre Lakatos*, Dordrecht 1976

Colbert, Jean-Baptiste (1619-1683) — 94
Colodny, R.S. (ed.) (a) *Beyond the Edge of Certainty*, Englewood
    Cliffs, N.J. 1965
  (ed.) (b) *The Nature and Function of Scientific Theories*,
    Pittsburgh, Penn. 1970
Copernicus, Nicolaus (1473-1543)
  (a)*On the Revolution of the Heavens* (A.M. Duncan, transl.),
    Newton Abbot 1976 (1543) (alternative transl.: *On the
    Revolutions* (E.Rosen, transl.), London 1978) — 9,36,38, 90,
    105,159,165,182-184,193,196,213,216,217,220,233,261,263-
    265,267,269,272,275,279,280,282,283,286
  (b)*Commentariolus* (c.1512), see Rosen — 36,38,191,193,217,
    220,261,263,272,273,281,284
  acceptance of his views — 103-107,161
  astrology — 215
  axiom — 11,98,101,132,193,194,217,222,223
  calendar (navigation) — 35,37,47,263
  coincidence (opposition, brightness) — 44,45,195,206
  counter-intuitive — 11,98,100,119,247,249,275
  cosmology (heliocentric, heliostatic, system) — 9,19,27-29,32,
    36,37,39,40,46,47,56,61,71,72,113,131,133,138,139,168,169,
    180,190, 193,195,207,217-223,244,258,267,280
  criticism — 37,116,120,206,212
  earth's motion — 11,19,36,37,44,47,61,63,72,83,90,96,169,184,
    190,191,222,223,245,247,258
  equant (deferent, epicycle) — 37,45,191,193,196,219-221
  explanation (prediction) — 31,38,41,46,52,56,62,222
  gravity — 44,71
  harmony (aesthetic) — 196,198
  heterocentric — 37,196,197
  hypothesis — 116,216,217,221-223,227,269,277,283
  instrumentalism — 36,51,64,192,217,220,282
  mathematics — 22,37,122,143,159,178,191,194,212
  measurement — 148,149,247,248
  order — 37,239,286
  parallax — 46,47,92,95,116,117,191,245,265
  polemic — 165
  problems — 95,96,107,138,207
  program(me) — 41,67,137,138,149
  relativity — 190
  retrograde motion — 38,45,47,48,66,88,92,137,167,169,191,195,
    280
  revolution — 9-11,26,84,103,107,109,151,171,244

simplicity (parsimony) — 189-191,193-195,281,282
size of planetary orbits — 47-49,88,195,207,265
space — 142,244,245
theology — 47,104,178,179,247
trepidation — 104,277
uniform circular motion — 40,44,100,106,132,137,149,222
Cotes, Roger (1682-1716) — 273,279
Coulomb, Charles Augustin de (1736-1806) — 79,147
Cusa, Nicholas of (1401-1464) — 35

Darwin, Charles Robert (1809-1882) — 89,182
Descartes, René (Cartesius) (1596-1650)
    Oevres (C. Adam, P. Tannery, eds.), Paris 1897-1913,
      reprinted 1964-1973 (12 vols.) — 284
    (a) *Discours de la Méthode*, avec *La Dioptrique, Les Météores,*
      La Géometrie, Oevres VI, Paris 1973 (1637) — 67,160,224,
      225,251,262,268,269,273,275,276,278,279,281,284-287
    (b) *Méditations touchant la Première Philosophie, Oevres*
      IX,I, Paris 1904 (1641) — 161,287
    (c) *Les Principes de la Philosophie, Oevres* IX,II, Paris 1904
      (1647) — 144,161,166,231,267-270,275,279,281,283-287
    (d) *Les Passions de l'Ame, Oevres* XI, Paris 1909 (1649) —
      283
    (e) *Le Monde, ou Traité de la Lumière, Oevres* XI, Paris 1909
      (1664) — 223,268,270,284,286,287
    action (by contact) — 72
    atoms (atomism) — 68,246
    church — 182,223,250
    circular motion — 67,70,72,100
    common sense (intuition) — 101,118,208,249,251,252
    double truth — 36
    doubt and certainty — 251
    experiment — 124,125,252
    explanation (motion by motion) — 41,52,67,69,73,83
    final cause — 243,250,286
    hypostatization — 83,204,249-252,259
    hypotheses — 225,226,252
    ideas (clear and distinct, evident) — 98-102,129,144,208,225,
      226,230,236,239,241,249,251,252,256
    impact (laws of) — 69,70,93,97,252,269
    inertia (law of) — 67,69,70,76,78,83,268
    laws of nature — 73,238,241,251,252
    magnetism — 68,71,73

mathematics — 143,144,148,155,166,208,224
matter (identical to extension) — 67,70,83,97,111,208,225,245,
    246,252,270
mechanism (mechanical philosophy) — 9,10,67,76,83,100,104,
    105,156,242,243,246,252
mind (and body) — 251,252,256,287
observation (sensory experience) — 68,251,252
occult qualities — 74
optics (light) — 68,151,224,225,252,285
planetary motion — 71,85,208,269
plenum — 71,137,225
primary (secondary) properties — 68,69
programme — 137
quantity of motion — 69,70,78,79,208,287
relativity (rest) — 70
space (place) — 67,70,245,246,286
vacuum (void) — 67,75,208,225
vortex (gravity) — 70,71,92,94,111,209,225,229,268,277,285
Dijksterhuis, E.J. (1892-1965) — 83,243
    De mechanisering van het wereldbeeld, Amsterdam 1950
        (*The Mechanization of the World Picture* (transl. C.
        Dikshoorn), London 1969 (1961)) — 261,263,265,267,270,
        272-275,277,281,284,286
Dooyeweerd, Herman (1894-1977) — 153,254,256,261,265
    A New Critique of Theoretical Thought (3 vols.), Amsterdam
        1953-1957 (1935-1936) — 254,277,287
Drake, Stillman (b.1910) — 160,199,265,281
    (ed.), *Discoveries and Opinions of Galileo*, Garden City, N.Y.
        1957 — 262,265-269,272,275,279,280,282,284,286
    (a) 'Copernicanism in Bruno, Kepler and Galileo,' in: Beer and
        Strand (eds.) 177-190 — 267
    (b) *Galileo at Work*, Chicago 1978 — 266-268,278-281,285
    (c) *Galileo Studies*, Ann Arbor 1970 — 267,268,270,275,279,280,
        282
Dreyer, J.L.E., *A History of Astronomy from Thales to Kepler*,
    New York 1953 (1906) — 261-263,266,273,279,284
Ducasse, C.J. — see Blake
Dürer, Albrecht (1471-1528) — 176
Duhem, Pierre (1861-1917) — 113
    (a) *The Aim and Structure of Physical Theory*, New York 1974
        (1906) — 274,284
    (b) *To save the Phenomena*, Chicago 1985 (1908) — 263,273,
        279,283,284

Einstein, Albert (1879-1954) — 70,77,80,82,92,127,174,190,194,278
Elkana, Y. (ed.), *The Interaction between Science and Philosophy*,
    Atlantic Highlands, N.J. 1974
Ellis, B., 'The Origin and Nature of Newton's Laws of Motion,' in:
    Colodny (ed.) (a) 29-68 — 270
Empedocles of Acragas (*c*.492-*c*.432 BC) — 53
Erasmus, Desiderius (1466-1536) — 176
Euclid of Alexandria (*c*.300 BC) — 54,99,142,144,245,265
Eudoxus of Cnidus (*c*.400-*c*.347 BC) — 32-34,206
Euler, Leonhard (1707-1783) — 137,148

Fermat, Pierre de (1601-1665) — 93,148,224,236,285
Fermi, L., Bernardini, G., *Galileo and the Scientific Revolution*,
    New York 1961 — 262
Feyerabend, Paul (b.1924) — 28,134,152,278
    (a) 'How to be a Good Empiricist,' in: Brody (ed.) 319-342 —
        262
    (b) 'Realism and Instrumentalism,' in: Bunge (ed.) 280-308 —
        262,273,281,282
    (c) 'Problems of Empiricism,' in: Colodny (ed.) (a) 145-260 —
        274
    (d) 'Problems of Empiricism II,' in: Colodny (ed.) (b) 275-353
        — 284,285
    (e) *Against Method*, London 1975 — 264,272,274,276,278
    (f) 'On the Critique of Scientific Reason,' in: Howson (ed.) 309-
        339, and in: R.S. Cohen *et al.* (eds.) 109-143 — 276
    (g) *Science in a Free Society*, London 1978 — 278
    see Cohen, R.S.
Finocchiaro, M.A. (a) *History of Science as Explanation*, Detroit
        1973 — 265,268,274,280
    (b) *Galileo and the Art of Reasoning*, Dordrecht 1980 — 261,
        262,264,268,276,278-280,282-285
Flamsteed, John (1646-1719) — 122,149,173
Foscarini, Paolo Antonio (1580-1616) — 179,180
Foucault, Jean Bernard Léon (1819-1868) — 82,285
Fracastoro, Girolamo (*c*.1478-1553) — 35,191
Freud, Sigmund (1856-1939) — 42

Galilei, Galileo (1564-1642)
    (a) *Dialogue Concerning the Two Chief World Systems* (S.
        Drake, transl.), Berkeley 1967 (1632) — 28,42,62,65,66,92,
        119,138,168-170,180,181,183,184,192,219,220,233,248,262-
        269,272, 275,277-286

(b) *Dialogues Concerning Two New Sciences* (H. Crew, A.de Salvio, transl.), New York 1954 (1638) — 28,38,63,107, 168,169,181,183,184,233,263,267,268,271,274-278,280-282, 285

Sidereus Nuncius (1610), *Letters on Sunspots* (1613), *Letter to Grand Duchess Christina* (1615), *Il Saggiatore* (1623), see Drake (ed.) — 161,166-168,179,180,262,265-270,272,279, 280,282,284,286

La Bilancetta (1586), see Fermi, Bernardini — 24

academy — 168,173

analogy — 141,277

anamnesis — 118,119

angular momentum — 40

astrology — 166,215

church (theology, Inquisition) — 36,61,64,66,104,168,170,178-184,219,223

circular motion — 62,63,67,72,100,138,198,199,282

comet — 87,166,167,180,198

common sense (intuition) — 119,207,247

cosmology — 50,62,166,168,169,178,207,219,236,244,245

criticism — 207

density — 23,24,65

discoveries — 13,61,93,106,150,176,279

earth's motion — 61-66,81,96,168,169,184,219

elements (atomism) — 60,62,163

empirical generalizations — 114,132

experiment — 65,124,125,150,164,248

explanation (motion by motion) — 41,52,63-66,73,83

floating bodies — 24,164

force — 74,76

gravity (fall) — 22,42,63-65,71,109-112,114,150,194,195,229, 274

harmony (aesthetic) — 198,199

hypotheses — 116,180,219,284

impact — 69

inertia (law of) — 62-64,66-68,83,198,207,247,282

instrumentalism — 64,192,219,220,234,267

Jesuits — 167-169,176,179-182

Jupiter's moons — 13,25,27,37,40,50,121,175,275

laws of nature — 73,238

light — 119,224,226

magnetism — 62,73

mathematics — 160,161,233,248,277

mechanics — 73,169,184,266,271
mechanism — 101
motion (and rest) — 62-65,70,235,267
natural motion — 63,64,71,198
pendulum — 81,109,124
phases (of Venus) — 13,96,218,220,284
Platonism — 64,67,75,104,119,143,161,233,239,246,285
popularization (didactics, education) — 160,161,164,165,169, 170
polemic — 24,166-170,193
primary (secondary) properties — 68
problems — 93,95,113,116,164
projectile motion (ballistics) — 64,109,124,141,266,274,275
publication (priority) — 183,214
relations — 22
relativity — 66,70,127
secondary light (of moon) — 96
sensory experience (observation) — 60,68,233
simplicity (parsimony) — 194,195
sound (music) — 64,199
space — 245
sunspots — 13,50,62,66,71,163,167,176,195,214,279
telescope — 12,93,121,150
tides — 65,66,73,141,168,180,268
time — 65,81,175
vacuum (void) — 63,65,246
Galilei, Vincenzio (1520-1591) — 199
Gassendi, Pierre (1592-1655) — 141,246
Giedymin, J., 'Instrumentalism and its Critique,' in: R.S. Cohen *et al.* (eds.) 179-207 — 262,263
Gilbert, William (1540-1603) — 73,104,107,176,247
    De Magnete (P.F. Mottelay, transl.), New York 1958 (1600) — 73,107,269
Gillispie, C.C., *The Edge of Objectivity*, Princeton, N.J. 1973 (1960) — 281
    (ed.), *Dictionary of Scientific Biography* (15 vols.), New York 1970-1976 — 261
Gimpel, J., *The Medieval Machine*, London 1977 — 280
Gingerich, O., "'Crisis' versus Aesthetic in the Copernican Revolution,' in: Beer and Strand (eds.) 85-93 — 273,281
Glymour, C., *Theory and Evidence*, Princeton, N.J. 1980 — 261,264, 274,276,277
Gödel, Kurt (b.1906) — 101

Goldbach, Christian (1690-1764) — 203
Goodfield, J. — see Toulmin
Grant, E., *Physical Science in the Middle Ages*, New York 1971 —
        280,285
Grassi, Orazio (1583-1658) — 167,176,279
Grünbaum, A. (a), *Philosophical Problems of Space and Time*,
        Dordrecht 1974 (1973) — 270,286
    (b) 'Is Falsifiability the Touchstone of Scientific
        Rationality?' in: R.S. Cohen *et al.* (eds.) 213-252 — 264,
        272,274
    (c) 'Popper versus Inductivism,' in: Radnitzky and Andersson
        (eds.) (a) 117-142 — 276
Guiducci, Mario (1585-1646) — 167,279
Guthrie, W.K.C., *A History of Greek Philosophy* (6 vols.),
        Cambridge 1980-1983 (1962-1981) — 266

Hacking, I., *The Emergence of Probability*, Cambridge 1975 — 286
Hall, A.R. (a) *The Scientific Revolution 1500-1800*, Boston 1966
        (1954) — 275
    (b) *From Galileo to Newton*, New York 1981 (1963) — 273
Halley, Edmund (1656-1742) — 145,173,269
Hanson, Norwood Russell (1924-1967) — 28,134,265
    (a) *Patterns of Discovery*, Cambridge 1958 — 262,268,270,276
    (b) 'Newton's First Law,' in: Colodny (ed.) (a) 6-a28,69-74
        —270
    (c) *Constellations and Conjectures*, Dordrecht 1973 — 264,265,
        269
Harman, P.M., *Metaphysics and Natural Philosophy*, Sussex 1982
        — 269,270
Heath, T.L., *Aristarchus of Samos*, New York 1981 (1913) — 261,
        262,266
Heilbron, J.L., *Electricity in the 17th and 18th Centuries*,
        Berkeley 1979 — 269,279,285
Heisenberg, Werner (1901-1976) — 174
Helden, A. van, 'Huygens and the Astronomers,' in: Bos *et al.*
        (eds.) 147-165 — 272,282
Helmholtz, Hermann von (1821-1894) — 80
Hempel, Carl G. (b.1905) — 46,205,229,274
    (a) *Fundamentals of Concept Formation in Empirical Science*,
        Chicago 1969 (1952) — 262
    (b) *Aspects of Scientific Explanation*, New York 1965 — 262,
        264,274,283
    (c) *Philosophy of Natural Science*, Englewood Cliffs, N.J.

1966 — 262,273-275

Heraclides Ponticus (*c.*390-after 339 BC) — 221

Hesse, M.B. (a) *Forces and Fields*, Totowa, N.J. 1965 (1961) — 278
   (b) *The Structure of Scientific Inference*, London 1974 — 274,
      286

Hipparchus of Rhodes (2nd century BC) —33

Hoeven, P. van der, *Metaphysica en fysica bij Descartes*,
      Gorinchem 1961 — 268,285

Hooke, Robert (1635-1703) — 131,141,145,173,213,214,224,227,269,
   270

Hooykaas, R. (a) *Geschiedenis der Natuurwetenschappen*,
      Utrecht 1971 — 263,264,285
   (b) *Religion and the Rise of Modern Science*, Edinburgh 1973
      (1972) — 275,280
   (c) *Capita Selecta*, Utrecht 1976 — 280,284

Horrocks, Jeremiah (1617-1641) — 269

Howson, C. (ed.), *Method and Appraisal in the Physical Sciences*,
      Cambridge 1976 — 276

Hübner, K., 'Descartes' Rules of Impact and their Criticism,' in:
      R.S. Cohen et al. (eds.) 299-310 — 269

Hume, David (1711-1776) — 187

Huygens, Christiaan (1629-1695) — 9,62,73,172,213,248
   Traité de la Lumière; Discours de la Cause de la Pesanteur,
      Brussels 1967 (1690) — 272
   académie — 94,173,177
   Cartesianism — 75,78,246
   chance — 93,236
   circular motion (centrifugal force) — 72,78,93,94,109,145,278
   discoveries —40,93,150,199
   force — 74
   gravity —71,94
   harmony — 199
   impact — 93,97,109,127,199,248
   inertia — 268
   inventions — 93,94,150,155,176
   light (waves) — 93,141,151,213,224,226,227,285
   mathematics — 93,143,148,155
   mechanism — 10,76,101,242,243
   normal science — 93,94
   pendulum (clock) —81,93,109
   quantity of motion — 76,93,97,269
   relativity — 70 97,127,199,248,268
   telescope — 93,150

Jammer, M. (a) *Concepts of Space*, New York 1960 (1954) — 286
(b) *Concepts of Force*, New York 19162 (1957) — 269,270
(c) *Concepts of Mass*, New York 1964 (1961) — 262
Jardine, N., *The Birth of History and Philosophy of Science*,
(Kepler's A defence of Tycho against Ursus), Cambridge 1984
— 284
Jordanus Nemorarius (before 1200-1237) — 74
Julius Caesar (100-44 BC) — 35

Kant, Immanuel (1724-1804) — 9,11,105,187,205,245,250,252,283
(a) *Kritik der reinen Vernunft*, Frankfurt a.M. 1974 (A: 1781,
B: 1787) — 11,261,268,270 273,276,283,286,287
(b) *Prolegomena zu einer jeden künftigen Metaphysik*,
Hamburg 1976 (1783) — 272
Kepler, Johannes (1571-1630)
(a) *Das Weltgeheimnis* (M. Caspar, transl.), München 1936
(1923) (1596,1621) — 49,97 131,164,263-265,272,276,282,
284,286
(b) *Neue Astronomie* (M. Caspar, transl.), München 1929
(1609) — 40,41,107,131,133,149,164,217,224,263,264,268,
269,276,278,280,282,284
(c) *Weltharmonik* (M. Caspar, transl.), München 1973 (1619)
— 40,53,197,264,266,282
Apologia Tychonis contra Ursum (1600-01), see Jardine — 217,
218,284
astrology (alchemy) — 74,197,215,243
didactic — 164
empirical generalizations — 39,41,115,131-133
force — 72-77,97
gravity — 71
harmony — 53,197-199
heliocentrism — 40,72
hypotheses — 217,218,227,284
instrumentalism (double truth) — 36,192,218
intuition — 101,240,247
Jesuits — 177
laws of nature — 102,218,238-241
magnetism — 71,73,74,143
mathematics — 122,248
model (polyhedra) — 49-51,62,88,131,197
motion (of Mars) — 39,50,71,72,90,114,133,149,191,217,248
observation — 75,100,114,131,149,150,156,240,248
optics — 121,224

parsimony — 194

pattern recognition — 115,131

planetary motion (laws of) — 34,39-42,48,50,72,75,79,86,97,
109-114,131,133,136,138,144-146,150,156,162,163,193,194,
197,208,220,227,238,240,247,269,277

rotation (influence) of sun — 71

satellite — 27,41,50

tides — 66,73

uniform circular motion — 39,40,71,72,97,100,106,131,198

Knowles, D., *The Evolution of Medieval Thought*, New York 1962
— 285

Koestler, Arthur (1905-1983) — 10,133

The Sleepwalkers, Harmondsworth 1972 (1959) — 261,263,
264,272,273,276,278,281-283

Kolakowski, L., *Positivist Philosophy*, Harmondsworth 1972
(1966) — 261,262,267

Koyré, A. (a) *Galileo Studies*, Hassocks 1978 (1939) — 263,265-
268,270,272,273,275,277,280,285

(b) *From the Closed World to the Infinite Universe*,
Baltimore 1974 (1957) — 286

(c) *The Astronomical Revolution*, Paris 1973 (1961) — 263-266,
268,269,273,275,276,278-284,286

(d) *Newtonian Studies*, Chicago 1968 (1965) — 268,269,271,
274,277,280,285,286

Kuhn, Thomas (b. 1922)

(a) *The Copernican Revolution*, Cambridge, Mass. 1971 (1957)
— 9,261,263,264,280,281,286

(b) *The Structure of Scientific Revolutions*, Chicago 1971
(1962) — 11,89,103,262,271-273,275,276,278

(c) 'The Function of Dogma in Scientific Research,' in: Brody
(ed.) 356-373 — 271

(d) *The Essential Tension*, Chicago 1977 — 280

anomaly — 92,93,103,105,272

crisis — 103,105,106,206

historical-relativism — 28

normal science — 89,91,94,102,202

paradigm — 89-92,134,157,250

problem — 89,91,92,102

revolution — 89,103

textbook — 91

theory-dependence (of statements, data, problems) — 28,91,92

Lactantius, Lucius C.F. (*c*.250-after 317) — 212

Lakatos, Imre (1922-1974) — 46,134-139,143,146,152,153,157,209, 272,276,277
  (a) *Proofs and Refutations*, Cambridge 1976 (1963-64) — 283
  (b) *The Methodology of Scientific Research Programmes*, Cambridge 1978 — 264,272,274-277,280,282-284
  (c) *Mathematics, Science and Epistemology*, Cambridge 1978 — 276
Lakatos, I., Musgrave, A. (eds.), *Criticism and the Growth of Knowledge*, Cambridge 1974 (1970)
Laudan, Larry — 134,157,235
  Progress and its Problems, Berkeley 1978 (1977) — 270,272-274,276,278,280,286
Lavoisier, Antoine Laurent (1743-1794) — 60
Leibniz, Gottfried Wilhelm (1646-1716) — 10,63,76,78,79,82,93,99, 101,127,143,148,160,190,194,213,214,223,242,246,262,270,287
Leonardo da Vinci (1452-1519) — 96,176,275
Lindberg, D.C. (ed.), *Science in the Middle Ages*, New York 1971 — 280
Louis XIV (1638-1715) — 94,173
Lovejoy, Arthur O., *The Great Chain of Being*, Cambridge Mass. 1976 (1936) — 286
Luther, Martin (1483-1546) — 103,177,178,239

Macey, S.L., *Clocks and the Cosmos*, Hamden, Conn. 1980 — 279, 280,286
Mach, Ernst (1838-1916) — 23,82,186-191,235
  The Science of Mechanics, (T.J. McCormack, transl.), La Salle, Ill. 1960 (1883) — 262,270,281,283
Madden, E.H. — see Blake
Maestlin, Michael (1550-1631) — 49,96,104
Margenau, H., *The Nature of Physical Reality*, New York 1950 — 281
Marx, Karl (1818-1883) — 42,175
Masterman, M., 'The Nature of a Paradigm,' in: Lakatos, Musgrave (eds.) 59-89 — 271
Maxwell, James Clerk (1831-1879) — 80,101,174
McMullin, E., *Newton on Matter and Activity*, Notre Dame 1978 — 270
Melanchton, Philipp (1497-1560) — 103,177,178
Mersenne, Marin (1588-1647) — 75,160,223
Michelangelo Buonarotti (1475-1564) — 176
Mises, R. von, *Positivism*, New York 1968 (1939) — 261,262,278
Motte, Andrew — 228,285

Müller, Johannes (Regiomontanus) (1436-1476) — 35,284
Musgrave, A., 'Evidential Support, Falsification, Heuristics, and
        Anarchism,' in: Radnitzky and Andersson (eds.) (a) 181-
        201 — 276
    see Lakatos

Nagel, E., *The Structure of Science*, New York 1961 — 270,273
    see Cohen, M.R.
Neugebauer, O. (a) *The Exact Sciences in Antiquity*, New York
        1969 (1957) — 263,281
    (b) *A History of Ancient Mathematical Astronomy* (3 vols.),
        Berlin 1975 — 263
Newton, Isaac (1642-1727)
    (a) *Philosophiae Naturalis Principia Mathematica* (A.
        Motte, E. Cajori, transl.), Berkeley 1971 (1687) — 23,76,
        90,98,105,109,111,114,131,138,143,144,159,160,165,166,
        183,185,195,208,209,212,214,226,227,231,246,262,268-274,
        277-279,282,283,285
    (b) *Opticks*, New York 1952 (1704) — 90,109,160,213,226,227,
        285
    (c) *Papers and Letters on Natural Philosophy* (I.B. Cohen,
        ed.), Cambridge, Mass. 1978 (1958) — 283,285
    action at a distance — 106,156,229
    alchemy (chemistry) — 213,243
    analogy — 141
    atomism — 24,80
    centripetal force — 77,78,144,145,278
    comet — 47,167
    conservation laws — 78-80,86
    crisis — 105,106
    criticism — 207,208
    density — 23,24,151
    dynamics (mechanics) — 73-80,82,91,108,140-142,145,162,184,
        270
    empiricism (observation, experiment) — 10,74,75,81,82,208,
        218,227,240,241,248
    force — 73-80,82,83,97,144
    force-matter dualism — 78,79,82,83,153,246
    generalization — 132,136,145,277
    gravity (law of) — 22,25,39,71,72,74-76,78,79,81,85,86,94,97,
        106,108-113,131,143-147,152,193,209,228-230,240,241,285
    hypotheses — 139,208,209,226-231,277,285
    induction — 228-230

inertia (inertial force) — 63,67,76,77,79,81,83,229
instrumentalism — 229,230
interaction — 83,140
intuition (common sense) — 98,99,101,166,208,230,247
laws of nature — 238,240
magnetism — 73,74,78,79,193
manifest (occult) properties — 74,156,229
mass (quantity of matter) — 23,24,40,69,76,79,145,147,151,229
mathematics — 76,78,85,86,106,108,143-148,162,166,194,208,
    214,248
mechanism — 9,74,76,83,208,229,242,243,285
model — 136
Newtonianism — 10,78,79,83,109
operational definitions — 23,69,76
optics (light) — 109,213,224,226,227,285
pendulum — 150
planetary motion (solar system) — 25,42,50,71,72,75,81,85,86,
    92,97,109,110,131,132,144-146,190,193,208,220
polemic — 165,166
primary (secondary) qualities (measurement) — 22,80,150,
    173,208
problems — 92,95,97,98
programme — 135,138
relativity — 77,78
rules of reasoning — 143-146,195
simplicity (parsimony) —193-195
space, time, motion — 70,76,77,79-82,127,245,246,286
synthesis — 10,105,109,138,274
three-body problem (motion of moon) — 193,194
tides — 66,97
vacuum (void) — 208,236,246,247
North, J., 'The Medieval Background to Copernicus,' in: Beer and
    Strand (eds.) 3-16 — 263

Ockham, William of (c.1300-c.1350) — 189,234
Oldenburg, Henry (c.1618-1677) — 173
Oresme, Nicole d' (c.1320-1382) — 35,36,96,263,277,279
Osiander, Andreas (Hossman) (1498-1552) — 51,177,192,216-218,
    220,284

Pacey, A., The Maze of Ingenuity, Cambridge, Mass. 1976 (1974) —
    278,280

Paracelsus, Philippus Theophrastus Bombast von Hohenheim
    (1493-1541) — 279
Parmenides of Elea (*c.*515-after 450 BC) — 57,60,266
Pascal, Blaise (1623-1662) — 10,24,75,93,98,109,141,148,151,172,
        182,213,235,236,240,246,248
    Oevres Complètes, Paris 1963 — 270,281
Pepys, Samuel (1633-1703) — 185
Peregrinus (Peter of Maricourt) (13th century) — 73
Peurbach, Georg (1423-1461) — 35
Plato (427-*c.*348 BC)
    The Collected Dialogues (E. Hamilton, H. Cairns, eds.),
        Princeton 1980 (1961) — 262,266,267,275
    academy — 171
    anamnesis — 118,119,247
    demiurge — 35
    elements (polyhedra) — 49,53,59
    ideas — 31,32,37,57,69,119,232-235,239,241,246,247
    influence — 67,68,143,233
    mathematics — 53,54,232,234
    motion (change) — 60,233-235
    observation (sensory experience) — 57,68,120,121,233
    Platonism — 9,10,37,45,60,62,64,75,104,161,165,233,285
    uniform circular motion — 31,32,95
Polanyi, M., *Personal Knowledge*, Chicago 1958 — 261,262,275
Popper, Karl R. (b.1902)
    (a) *The Logic of Scientific Discovery*, New York 1968
        (1934,1959) — 11,261,264,271-276,283,286
    (b) *Conjectures and Refutations*, London 1976 (1963) — 261,262,
        264,266,271-273,275,276,278,282-284,286
    (c) 'Normal Science and its Dangers,' in: Lakatos, Musgrave
        (eds.) 51-58 — 272
    (d) *Objective Knowledge*, Oxford 1974 (1972) — 261,272-274,
        276,283,286
    (e) 'Autobiography,' in: Schilpp (ed.) 1-181 (also published
        as *Unended Quest*, 1976) — 264
    (f) *Realism and the Aim of Science*, London 1983 — 261,264,
        276,282,283,286,287
    ad-hoc hypotheses (conventionalist stratagem) — 114-116
    basic statements — 115,274
    corroboration — 205,237
    criticism — 200-205,237,252-255
    demarcation — 43,202,203,255
    falsification — 42,43,113-115,130,202-204,237,253,274

hypostatization — 201,202,204,249,255
induction (inductivism, deduction) — 113,129-133,152,201,202, 253
inertial (gravitational) mass — 147
instrumentalism — 192,237,253,282
laws of nature — 253,254,264
mathematics — 43,143,203,204,255
meaning (truth, realism) — 26,43,202,204,234,240,253
normal science — 91,202,272
paradox — 204,205
physicalism — 204,249
problems — 89
progress — 152,200
theories — 14,89,261,271,274
transcendence (metaphysics) — 43,253-256
transcendental critique — 252-255
trial and error (conjectures and refutations, hypothetical-deductive method) — 89,114-116,131,132,147,152,202,229, 249,253,255,275
vulnerability — 114
worlds — 14,133,201,204,252-254,260
Price, D.J. de S., *Science since Babylon,* New Haven 1975 (1961) — 278,280
Ptolemy (Ptolemaios) of Alexandria, Claudius (*c*.100-*c*.170)
Almagest — 33,34,39,90
astrology — 34
astronomy and physics — 34,35
deferents (epicycles, equants, excenters, inequalities) — 33-35, 37,39,193,196,219-221
heterocentric circles — 34,169,197,219,220
instrumentalism — 34,192
order of planets — 47,48,195,196,265
retrograde motion — 33,44,45,92,195,264
rotation of earth — 96
system — 27,38-40,45-47,51,61,90,105,106,113,117,133,138, 148,169,179,181,189-192,194,195,206,207,217,218,220-222, 282,284
uniform circular motion — 34,44,222
Pythagoras of Samos (*c*.560-*c*.480 BC) — 10,19,20,52-54,61,104,122, 140,143,196,199,232,266,273

Radnitzky, G., 'Justifying a Theory versus Giving Good Reasons for Preferring a Theory,' in: Radnitzky and Andersson (eds.) (b)

213-256 — 265

Radnitzky, G., Andersson, G. (eds.) (a) *Progress and Rationality in Science*, Dordrecht 1978
   (b) *The Structure and Development of Science*, Dordrecht 1979

Ramus, Petrus (Pierre de la Ramée) (1515-1572) — 104,164,175,177, 217,218

Randall, J.H., *Aristotle*, New York 1960 — 262,276,287

Ravetz, J.R., *Scientific Knowledge and its Social Problems*, Harmondsworth 1973 (1971) — 261,270,271,276

Regiomontanus — see Müller

Reichenbach, H. (a) *Experience and Prediction*, Chicago 1970 (1938) — 274,276,286
   (b) *The Rise of Scientific Philosophy*, Berkeley 1968 (1951) — 276

Reinhold, Erasmus (1511-1553) — 105,178

Rescher, N., 'Some Issues Regarding the Completeness of Science and the Limits of Scientific Knowledge,' in: Radnitzky and Andersson (eds.) (b) 19-40 — 272

Rheticus, Georg Joachim (1514-1576) — 36,177,178,196,217,284
   Narratio Prima (1540) — see Rosen (a) — 36,71,263,265,282

Richer, Jean (c.1630-1696) — 151

Rohault, Jacques (1620-1675) — 161

Rømer, Ole (1644-1701) — 151,173,226,236

Ronchi, V., *The Nature of Light*, London 1970 (1939) — 284

Rosen, E., (a) *Three Copernican Treatises*, New York 1959 (1939) — 261,263,265,272,281,282,284
   (b) *Copernicus and the Scientific Revolution*, Malabar, Fl. 1984 — 263,273,284

Rudolph II of Habsburg (1552-1612) — 149,172,177,217

Rudwick, M.J.S. — see Bos

Russell, B., *History of Western Philosophy*, London 1961 (1946) — 285

Sabra, A.I., *Theories of Light from Descartes to Newton*, Cambridge 1981 (1967) — 284,285

Sagredo, Giofrancesco (1571-1620) — 29,62,168,279

Salmon, W.C. (ed.), *Zeno's Paradoxes*, Indianapolis 1970 — 266

Salviati, Filippo (1582-1614) — 29,42,164,168,279,280

Santillana, G. de, *The Crime of Galileo*, Chicago 1976 (1955) — 281

Scheiner, Christoph (1573-1650) — 167,176,214

Schilpp, P.A. (ed.), *The Philosophy of K.R. Popper*, La Salle, Ill. 1974 — 276

Scott, J.F., *The Scientific Work of René Descartes*, London 1976
(1952) — 268,284

Shea, W.R., *Galileo's Intellectual Revolution*, London 1972 — 268,
275,279

Simon, H.A., *Models of Discovery*, Dordrecht 1977 — 276

Simplicius (*c*.500-after 533) (Simplicio) — 29,42,138,164,168,169,
279,280

Sneed, J.D., *The Logical Structure of Mathematical Physics*,
Dordrecht 1979 (1971) — 271

Snelders, H.A.M. — see Bos

Snel(lius), Willebrord (Snel van Royen) (1591-1626) — 188,213,225

Socrates of Athens (*c*.470-399 BC) — 24,119

Sophocles (*c*.496-*c*.405 BC) — 165

Stafleu, M.D. (a) *Time and Again, A Systematic Analysis of the
Foundations of Physics*, Toronto 1980 — 266,275,278

(b) 'The Mathematical and the Technical Opening-up of a
Field of Science,' *Philosophia Reformata* 43 (1978) 18-37
— 270,278

(c) 'The Isolation of a Field of Science,' *Phil. Ref.* 44 (1979) 1-
15 — 278

(d) 'The Opening-up of a Field of Science by Abstraction and
Synthesis,' *Phil. Ref.* 45 (1980) 47-76 — 270,278

(e) 'Theories as Logically Qualified Artefacts,' *Phil. Ref.* 46
(1981) 164-189; 47 (1982) 20-40 — 261

Stegmüller, W., *The Structuralist View of Theories*, Berlin 1979 —
278

Stevin, Simon (1548-1620) — 74,75,95,104,124,160,176,199

Stokes, George Gabriel (1819-1903) — 266

Strand, K.A. — see Beer, A.

Suppe, F., 'The Search for Philosophic Understanding of
Scientific Theories'; 'Afterword — 1977'; in: Suppe (ed.) 3-
241; 617-730 — 261,271,276-278

Suppe, F. (ed.), *The Structure of Scientific Theories*, Urbana, Ill.
1979 (1973)

Suppes, P., *Introduction to Logic*, New York 1957 — 261,262

Szabo, I., *Geschichte der mechanischen Prinzipien*, Basel 1977 —
270

Tartaglia, Nicolo (*c*.1505-1557) — 37,74,176,267

Tertullian, Quintus Septimus Florens (*c*.160-after 220) — 280

Thomas Aquinas (1225-1274) — 35,99,178,233,234

Torricelli, Evangelista (1608-1647) — 24,74,75,93,109,141,151,246,
279

Toulmin, S.E. (a) *Foresight and Understanding*, London 1961 —
    265,284
    (b) *Human Understanding*, Princeton 1977 (1972) — 271
Toulmin, S., Goodfield, J., *The Fabric of the Heavens*, London 1961
    — 261,263,266,281
Trismegistus (Hermes) — 165
Tycho — see Brahe

Urbach, P., 'The Objective Promise of a Research Programme,' in:
    Radnitzky and Andersson (eds.) (a) 99-113 — 276
Urban VIII — see Barberini, M.
Ursus — see Bär

Velikovski, Immanuel (1895-1979) — 281
Vinci — see Leonardo
Visser, R.P.W. — see Bos
Voltaire, François-Marie (Arouet) (1694-1778) — 161,247

Wallis, John (1616-1703) — 97,109
Wartofsky, M.W. — see Cohen, R.S.
Watkins, J., 'Against Normal Science,' in: Lakatos, Musgrave
    (eds.) 25-37 — 272
Weimer, W.B., *Notes on the Methodology of Scientific Research*,
    Hillsdale, N.J. 1979 — 261
Westfall, R.S., *Never at Rest, A Biography of Isaac Newton*,
    Cambridge 1980 — 268,270,272,275-277,286
White, A.D. (1832-1918) — 280
Whittaker, E.T., *A History of the Theories of Aether and
    Electricity*, vol. I, New York 1973 (1910 — 284
Wightman, W.P.D., *Science in a Renaissance Society*, London 1972
    — 280
Worrall, J., 'The Way in which the Methodology of Scientific
    Research Programmes improves on Popper's Methodology,' in:
    Radnitzky and Andersson (eds.) (a) 45-70 — 276
Wren, Christopher (1632-1723) — 97,109,145,269

Zahar, E., 'Why did Einstein's Programme supersede Lorentz's?'
    in: Howson (ed.) 211-275 — 276
Zeno of Elea (*c.*490-*c.*425 BC) — 57,60,61,266,267
Zilsel, E., *Die soziale Ursprünge der neuzeitlichen Wissenschaft*,
    Frankfurt a.M. 1976 — 280
Ziman, J.M. (a) *Public Knowledge*, Cambridge 1974 (1968) — 279,
    281,283

(b) *The Force of Knowledge*, Cambridge 1976 — 277,279,280
(c) *Reliable Knowledge*, Cambridge 1978 — 276,279